WEATHER WATCH

How to Make the Most of America's Changing Climate

HAROLD W. BERNARD, Jr.

Foreword by John Coleman

WALKER AND COMPANY   *New York*

First published in the United States of America in 1979 by the Walker Publishing Company, Inc.

Published simultaneously in Canada by Beaverbooks, Limited, Pickering, Ontario.

ISBN: 0-8027-0608-8

Library of Congress Catalog Card Number: 78-57686

Printed in the United States of America

10 9 8 7 6 5 4 3 2

The opinions, findings, and conclusions in this book are those of the author alone (except as noted) and do not necessarily reflect those of any of the author's past or present affiliations.

To My Wife, Christina,
  Who Said I Could;
To My Mother, Evelyn,
  Who Knew I Would;
To My Father, Harold,
  Who Led the Way
  (But Didn't Tell Me How Much
  Work It Was to Follow)

Contents

ACKNOWLEDGMENTS

This book would not have been written without the theories and ideas of Dr. Hurd C. Willett, professor emeritus of Meteorology at MIT and now president of The Solar Climatic Research Institute, Inc. I spent more than a few hours with Dr. Willett and found him as energetic physically as mentally: Even though he is almost four decades my senior, I was hard-pressed to keep pace with him on a cross-country ski journey through the Vermont woods. I remember the skis he wore appeared ancient—nothing more than wide, flat boards.

But sometimes the old equipment is the most reliable and the most useful. So may it be with ideas. It is just that sometimes we have to chase them through the snow for a while to find out. Dr. Willett's ideas have not been widely accepted, but evidence is mounting now that he may have been on the right track all along. And he has not yet given up skiing.

Bob Clawson, of Impact Advertising, Inc., critiqued and encouraged my early writing efforts. I figured if I had an ex–English teacher in my corner, I could do the job.

Ernie Abrams, a meteorologist at Environmental Research & Technology, Inc. (ERT), and a retired Air Force weather officer, reviewed all of the chapters I wrote, and offered a number of helpful comments. Ernie had an especially difficult job because the chapters kept coming to him out of order and without the graphics.

Susanah H. Michener (ERT) translated my scrawlings, which I

called graphics, into some fine illustrations for the book; and Carol Maneely was able to decipher my rambling, cut-and-paste drafts and produce a professionally typed manuscript. Carol, it turned out, had a fine editorial eye, too.

But so did Walker and Company's Richard Winslow, and I thank him for getting me through this project.

My thanks to the following people who took time out to answer questions and requests for information: H. Prescott Sleeper, Jr., Kentron International; Robert Lautzenheiser, New England Climatic Service; David M. Ludlum, long-time editor of *Weatherwise*; David M. Meko, University of Arizona; Charles M. Umpenhauer, KOA-TV; Dr. James McQuigg, former director of the U.S. Center for Climatic and Environmental Assessment; Peter R. Leavitt, Weather Services Corporation; Patrick S. McIntosh, National Oceanic & Atmospheric Administration (NOAA) Space Environment Services Center; Earl S. Finckle, Central Weather Service; Paul E. Damon, University of Arizona; and Eric S. Dickman, The Energy Efficiency Corporation.

And finally my thanks to Tara Taylor, who did lots of calculations on her Wise Old Owl Calculator but didn't get any answers. At least she was quiet.

Foreword

"The Ice Age is coming! The Ice Age is coming!," intoned the modern Paul Revere as he rode the airwaves of talk shows from coast to coast. And the cover of his book, emblazened with a thousand-foot-high sheet of ice, was highlighted in bookstores in every part of the country. His picture didn't make the cover of *Time* magazine, but dollar bills began to accumulate in his bank account as shivers of frozen death swept the nation.

This is not that book.

In fact, that book has not, to the best of my knowledge, been published yet. But a whole barrage of books telling of disastrous climatic change can be found. Mostly, they are gleaned from the press releases issued in connection with publication of scientific papers and clippings gathered from press interviews with scientists. By weaving such materials together with drama and little regard for proper balance, emphasis, or perspective, a rather devastating picture has been painted.

This is not one of those books, either.

Harold "Buzz" Bernard has taken a much more reasonable, studied, and personal approach. The results do foretell of an important change in our lifestyles, economy, and perhaps history, forced upon us by an altered climate. And his projections do extend into an "Ice Age" if you are willing to use the broad, not the comic book, definition.

Discounting a "snow blitz" and the havoc created by a sudden shift of the North Pole to Philadelphia, and instead digging deep for scientific precedent and charting the long-range effects, "Buzz" takes us carefully through what should be the climatic conditions for the balance of our lives (if his basic theories and assumptions are correct). He presents his findings with enough support to let you calculate the effects on your life and business, and consistently presents a text that is clear and straightforward to the non-scientific reader.

A nice job, well done, and probably correct; but I can no more vouch for his forecast than he could for one of mine on "Good Morning America."

In fact, as a man who struggles most of the night in an often vain effort to get today's forecast for the nation right, I am somewhat awed by the task of projecting the next 20 to 30 years of climate. One nice thing about television—most people forget what I said before tomorrow morning. I wonder if "Buzz" realizes his book will be sitting on shelves 3 or 4 decades from now and that those who care can analyze in detail just how accurate his outlook was. If it goes bad, at least he'll have time to change his name and grow a beard.

                                        John Coleman
                                "Good Morning America"
                                        ABC TV

Preface

THIS BOOK is not about an imminent ice age, the impending flooding of our coastal cities, or the inevitable desiccation of the Midwest. Doomsday books make for headlines and exciting reading. It seems we all love to be scared to death, whether it be by monsters or ice ages.

There is no monster in this book. But there is a gremlin who is going to steal more of your income and mine, too. The gremlin is climate change. The weather of the next several decades is going to be somewhat different from that of the past few decades. The differences are going to cost us money, and perhaps dictate some changes in our life-styles.

The differences will not be large in an absolute sense, but they will be significant. I will be talking about changes of 1 or 2 degrees (Fahrenheit*) in many cases. If that doesn't sound like much, consider them against a background where a change of 6 or 7 degrees, on a global scale, would bring us to the threshold of an ice age.

An ice age is not lurking in the shadows of the near future. But colder, snowier winters probably are, at least for a good part of the United States. Wetter conditions, on the average, should prevail in the immediate future, too. The specter of major drought may be lying in ambush behind the wetter weather, though. And next time it may not be the Midwest that will be fighting off the threat of a Dust Bowl.

*Fahrenheit will be used throughout this book, except where specifically noted.

Such violent weather phenomena as tornadoes and hurricanes are part of the changing climate. Over the next twenty or thirty years hurricanes may follow a more "traditional" track than they did during the 1940s, 1950s, and 1960s. But major tornado outbreaks may threaten areas outside the "traditional" tornado alley.

What I offer in this book are not specific forecasts for specific events to happen at specific times. Such forecasts have no scientific validity. What I do offer are outlooks for average conditions over the next 30 years, and a comparison of how (and how much) those conditions might differ from what we have come to consider as "normal." I will also suggest how the changing conditions may affect our pocketbooks and our way of living.

Admittedly, the book has a biased viewpoint. But it is biased only because I have taken a position on what I think might happen. I believe it is the most logical position, but you might not agree. For that reason I have spelled out in detail in the book how I arrived at the conclusions I did.

Getting at those conclusions took a few years, and the help of others. But it was exciting work, and I believe it makes for fascinating reading.

One final note: I have mentioned weather extremes (all-time records) a number of places in the book. I tried to keep updated on those extremes as I prepared the manuscript, but new records tumbled even as the book was being finished. I therefore apologize if some of the records mentioned are already outdated. The winter of 1977-78 was particularly rough on weather records. So was the winter of 1976-77 for that matter.

It is interesting to note that the winter of 1977-78, with its copious rain and snow, was perhaps much closer to the type of winter we will see more of in the near future than was the previous winter (1976-77) with its bitter cold in the East and drought in the West. The winter of 1977-78 was perhaps an extreme example of what is coming up—but why do you not read on and judge for yourself?

CHAPTER 1

On the Brink

THERE IS general agreement among those who study climate and meteorology that the earth is on the brink of a significant climatic trend. The weather of the mid-1950s to the mid-1970s was uncommonly kind to us in the United States.[1] But that's coming to an end now. The winters of 1976-77 and 1977-78 were two of the harshest in recent memory. We have been warned by a senior research climatologist to expect "greater droughts, more floods."[2]

However, beyond the point of consensus that the climate of the immediate future will be less benign, there is little if any agreement as to which way the climatic trend will point—colder or warmer. That, obviously, is the more important consideration. The weather of the coming decades will affect all of us. It will, in part, determine our cost of living, how much we will have to shell out for food and fuel and clothing. It will affect our life-style, perhaps dictating, for instance, the need for water rationing in some parts of the country, and suggesting we develop a greater appreciation for winter sports in others. It will affect our health, since it will influence such things as our endurance and productivity. Indeed, it may affect our longevity.

A trend toward colder weather would increase demand for such things as heating fuel and heavier clothing. More demand: higher prices. Colder weather could lead to greater snowfall. That in turn might boost sales of snow tires and front-wheel-drive cars. Along with deeper snows would come the increased costs of snow removal and the economic penalties of disrupted commerce. Food prices, for example, might jump in response to the difficulties associated with transporting food to market through ice and snow.

Colder weather would foster a boom in winter sports, such as skiing and ice skating. But it might have a deleterious effect on winter tourist industries in Florida and Arizona if the traditionally balmy January temperatures in those states fell prey to more frequent cold spells. On the other hand, chillier temperatures could accelerate the recent permanent population migration into the Sunbelt regions of the South as people flee the rigors of repeated severe northern winters. Such southern regions might see their attraction as retirement havens grow even stronger.

Finally, the implications of a cooler climate extend even to our personal health: Lower temperatures could help us to live longer. Death rates are generally higher in warmer areas of the country than they are in colder ones.

A warming trend might have just the opposite effect, though. And although warmer weather would not force us to buy heavier clothing and spend more money to keep our houses warm, it could overburden electric utilities with a skyrocketing demand for power to run air conditioners.

With higher temperatures usually comes the threat of drought, too. In some parts of the United States, a severe drought would strain water supplies, which in turn would dictate the need for water rationing: Crops would suffer and food prices would soar. In the extreme, a super-drought might bring about a repeat of the tragedy of the 1930s, in which the Great Plains became a great dust bowl. People by the thousands streamed out of the Great Plains states then, abandoning the region to dust and wind.

The potential climate-related problems extend well beyond our personal lives and the United States, of course. The most basic worldwide problem is food. But it is not the intent of this book to address the economic, social, and political impediments associated with feeding the world's population. That has already been done comprehensively, and quite eloquently, by Dr. Stephen H. Schneider in his book *The Genesis Strategy.*[3]

It *is* the intent of this book to set forth a series of specific climatic projections for the United States, to note how the projected conditions would differ from those of the recent past, and to examine ways in which the differences may affect us. A special chapter, chapter 3, is devoted to more severe weather, such as hurricanes, tornadoes, and drought. The area of maximum threat for those important phenomena will likely change from what we have experienced over the past 20 or 30 years. The chapter also discusses ways in which we can react to those threats.

Of course, the theories and viewpoints relating to the topic of climate change are myriad. There are as many outlooks as there are meteorologists and climatologists. It is fair to ask, then, how I can pretend to know what is going to happen. I do not know for sure. No one does. But I do have a strong belief, and plenty of persuasive scientific data to back it up, that one particular climatic course is much more likely than any other.

In general, I expect cooler summers and colder winters to plague much of the United States over the next several decades. Many areas should be wetter, a few places drier, some snowier, but in no part of the United States should it be significantly warmer. And lurking behind the dreariness of increased rainfall will be the specter of drought, at least for one particular area of the country. What this all means in terms of such things as heating bills, storm preparations, and severity of winters is discussed, region by region, in the following chapters. The chapter, chapter 15, on personal implications, discusses changes in our life-style likely to result from a trend toward cooler weather.

The climate over the next quarter century will be different from what we've become used to. There will be changes in rainfall, snowfall, temperature, and cloudiness. The challenge is to try to quantify those changes. Rather than try to tell you that an ice age is imminent (it is not), or warn you that the Midwest is going to blow away in a series of choking duststorms (it will not), or suggest that New York and other coastal cities will drown under the influence of melting polar ice caps (they will not), I would rather set down in numbers and narratives a more realistic perspective on what to expect.

I may be entirely wrong in my expectations, and I will be roundly criticized by some of my colleagues for even writing about such speculative matters. But I would rather be wrong and be criticized, in attempting to present what I feel is a rational approach to climate projection, than continue to listen to vague, general statements about impending change, and in some cases doom.

My approach, as amplified in the next chapter, is to argue that the next 2 or 3 decades will experience climatic conditions similar to those of the early 1800s. Unfortunately, the only region in the United States for which comprehensive and complete weather records exist for that era is New England. That in itself makes an interesting study and comparison with the immediate past (see chapter 12), but it does not necessarily reflect nationwide conditions. So, in lieu of the early-1800s, I examined weather records from the late 1800s and early 1900s, a period similar to, but not quite so extreme as the early nineteenth century. Again, this is explained in the next chapter.

I chose twenty-four cities to study (figure 1). The cities were chosen for their various geographical locations as well as for the length of their records. For each city I calculated averages of temperature, precipitation, snowfall, and sunshine for the period 1880–1909. I selected a 30-year period because these turn-of-the-century averages or normals are compared with what the National Weather Service (NWS) defines as contemporary normals, which

1. The twenty-four key cities chosen to study. The cities were chosen for their geographical locations as well as for the length of their weather records.

are the averages for the 30-year period 1941–1970. When you hear your local meteorologist say that last month was 3 degrees below normal, he is comparing it with the 30-year average from 1941 through 1970. New normals will be defined after 1980. They will be the averages from 1951 through 1980.

In the appropriate regional chapter I include a table for each city studied. The tables allow you to compare the current normals in a particular region to those which I speculate will be closer to the normals of the next few decades, as represented by the period 1880–1909. You will see that there are some important changes to be expected in some areas or some seasons, whereas in other seasons the climate may not be much different 20 years from now than it was 20 years ago.

Let me emphasize that the climate forecasts I present are speculative. They are based on analogies with the past. But given the fact we do not know for sure what causes short-term climate changes, I would argue that such an approach to prediction is the most valid.

Lord Byron said it in 1821: "The best of prophets of the future is the past."

Notes

1. Bernard, H. and Spiegler, D. "The Winter of 1976–77—A Mid–season Perspective, *Enviromap Newsletter*, January 1977, p. 4.
2. Pierce, R. (Newhouse News Service) in an article in the *Boston Evening Globe*, 10 November 1976, quoting J. Murray Mitchell, Jr., Senior Research Climatologist, U.S. Environmental Data Service.
3. Schneider, S. and Mesirow, L. *The Genesis Strategy* (New York: Plenum Publishing Corp., 1976).

CHAPTER 2

Cycles, Ice Cores, and Sunspots

CERTAIN CLIMATIC CYCLES are known to have been in existence for anywhere over the past 200 to 800 years. Admittedly that is but a brief instant in the history of the earth, but 800 years is a relatively long time when you consider we are concerned with only the next 20 to 30 years.

Because comprehensive, written records of climate have been kept for only about 100 years or so, at least in the United States, it is perhaps a little difficult to understand how scientists could discover a cycle that goes back eight hundred years, or even two hundred. Climate records do not necessarily have to be compiled by human observation or direct measurement of temperature and precipitation. They can be etched in natural "logbooks," such as layers of Greenland ice, where they can accumulate for centuries. By drawing ice core samples from the Greenland ice cap and studying the amount of an oxygen isotope—an isotope is one of the variant forms of a single element—in various layers of those cores, researchers can determine which historical eras were relatively cold and which were relatively mild.

Old nonmeteorological records can be examined for certain other climatic clues. For instance, historical observations of the advance and retreat of European glaciers can tell climatologists something about the warmness or coldness of times past. Notations on the number of months that certain North Atlantic coasts were icebound can hint at different climatic regimes. And records of the rise and fall of colonies on Greenland may reflect changing climate.

Although cause and effect have not yet been established, there are some strong correlations between the climatic cycles that researchers have uncovered and solar activity. Solar activity may best be indicated by the number of spots on the surface of the sun—that is, sunspots. Fortunately, there's a pretty good record of sunspot numbers going back to the year 1610.[1]

Supporting the sunspot-climate theories are modern mathematical calculations and ancient information stored in the annual growth rings of trees. The information locked in the tree rings is in the form of a radioactive carbon isotope. The amount, or concentration, of the isotope is proportional to the degree of solar activity that was occurring at the time the isotope was absorbed by the tree. In oversimplified terms, tree ring researchers can estimate the amount of solar activity that occurred in a particular year by measuring the concentration of the carbon isotope in the tree ring corresponding to that year. The specific year can be determined simply by counting the number of growth rings between the ring being studied and a ring of known date.

By correlating the mathematical calculations and the carbon data to the climate records of ice cores and various historical documents, and by projecting the resulting relationships forward in time, we can get a pretty good idea of what lies in our immediate climatic future.

We must make one rather important assumption to be able to do this, though. That assumption is that the cycles we are concerned with will continue and will not be altered or overridden by any stronger climatic trend, such as that toward an ice age, or in the other direction, toward unprecedented warming. Such trends could be either man-caused (*anthropogenic*, as scientists like to say) or natural. However, there is no convincing evidence that any such stronger climatic trend is underway. Therefore, the assumption that the cycles will continue is reasonable.

Man-caused influences are becoming more important, however, and may indeed be able to dominate the natural cycles by early next century. I talk more about that in chapter 14. But let us first examine where our climate is heading over the next several decades.

Ice Cores

Ice core samples from Greenland may offer a major clue as to what the climatic trend of the near future will be. In 1966 the U.S. Army Cold Region Research and Engineering Laboratory, using a modified oil-drilling rig, extracted a 4,560-foot-long ice core from the northern Greenland ice sheet. The core contained a climatic history of at least 100,000 years. Climatologists from all over the world have studied samples of that core.

What many of them were particularly interested in was the amount of an oxygen isotope in various layers of the core. The isotope of interest was oxygen 18, which is a slightly heavier variant—on the atomic weight scale—of the common form of oxygen, oxygen 16.

Oxygen is a constituent of water, of course. The snow that falls on Greenland is formed from water evaporated from surrounding oceans. Water molecules containing the lighter oxygen 16 evaporate more readily than ones with the heavier isotope; they also remain evaporated longer. The minority of molecules containing the heavier oxygen 18 evaporate less easily, and they remain evaporated—that is, in a vapor state—for a shorter period.

When the air is colder, evaporation is even slower and precipitation out of the vapor state is faster. Or, saying it another way, in a colder atmosphere fewer of the molecules containing the heavier isotope evaporate; and those that do evaporate tend to change more rapidly from a vapor state, falling as rain or snow before they can reach Greenland. The molecules that do make it to Greenland are deposited in snowfall there, and the snow is eventually compacted into ice.

By studying the varying amounts of oxygen 18 in different ice samples, scientists can thus tell whether the climate was colder or warmer at the time the ice was formed. The less oxygen 18 present, the colder the climate was. The age of an ice sample can be calculated from its depth within the core. The actual measurement of oxygen 18 is done by a machine called a mass spectrometer, which sorts molecules according to their atomic weights.

Ice core samples examined by Danish scientist Willi Dansgaard show that significant cooling periods occurred roughly every 80 and 180 years.[2] The 180-year cycle, as we shall see, is the more im-

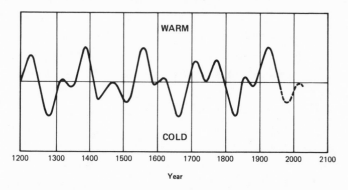

2. A schematic depiction of global temperatures as suggested by studies of Greenland ice core samples. The implication is that large-scale cooling takes place roughly every 180 years. If that cycle continues, the 1980s will bring the next marked drop in worldwide temperatures. (Dr. Wallace S. Broecker, *Science*, vol. 189, pp. 460–463, fig. 2, August 8, 1975. Copyright 1975 by the American Association for the Advancement of Science.)

portant. Figure 2 presents a schematic view of the cycles over the past 800 years. If the cycles are projected into the future, they suggest that the 1980s will bring a marked drop in global temperature, the manifestation of the 180-year cycle.[3] That cycle in the past has coincided with such events as the disappearance of the Norse colonies on Greenland, the peak of the "Little Ice Age" in Europe, and a period in which occurred New England's "Year Without a Summer," 1816.

According to the ice core samples, a stretch of cold years reached a nadir in the mid-1400s. It was shortly after that that the colonies on Greenland, established by Norsemen during an unusually mild spell between the years 800 and 1000, vanished. The disappearance was at least partially due to a deteriorating climate, probably more persistent ice and snow.[4]

The samples indicate another cold term reached its minimum in the early 1600s. The culmination of the "Little Ice Age" came during the 1600s. Alpine glaciers expanded to their maximum extents in historical times, burying pasturelands and passes in great sheets of ice; the shores of Iceland, now ice-free except for a few weeks each year, were locked in ice for up to 6 months every year.[5]

The most recent major cold regime evidenced in the ice core was centered around 1800. The "Year Without a Summer" was 1816. Snow whitened parts of the New England countryside in June; light frosts struck in July and August, then heavier frosts destroyed the corn crop in September. It was the coldest summer on record in New Haven, Connecticut, and Hoheneissenberg, Bavaria.[6]

All in all, the ice core evidence for significant global cooling every 180 years is rather substantial and points toward a recurrence in the 1980s. By the early 1970s, Professor Hubert H. Lamb of the Climatic Research Unit in England may already have detected the beginning of such cooling in the high-altitude wind patterns of the Northern Hemisphere: "Increasingly they resemble the wind patterns of the 'Little Ice Age.' "[7]

Professor Lamb was able to determine the configuration of up-per winds associated with the "Little Ice Age" by studying the sparse records of surface temperature and atmospheric pressure available from that era. By applying a deductive process involving knowledge of the various vertical relationships among tem-perature, pressure, and wind, he was able to reconstruct what the probable high-level wind patterns of that time were.[8]

The Not-So-Constant Solar Constant

The sun has long been considered by most of us to be a fixed, un-varying entity in our universe, steadily pouring out warmth to us day after day. We take its consistency for granted.

Perhaps we shouldn't. A 1976 article in *Science* stated, "Indeed, what is now being suggested is that the sun, far from being the constant star of recent memory and astronomical theory, has in the past 1,000 years undergone several significant changes in its magnetic activity and, perhaps, in its output of energy. If so, then future changes in solar activity cannot be ruled out—a prospect with profound implications for solar physics and, possibly, for the earth's climate."[9]

A number of years ago a group of scientists headed by C. G. Abbot of the Smithsonian Institution, and more recently a number of Soviet researchers, theorized that the *solar constant* (a number representing the radiative output of the sun) varies with the amount of sunspot activity.[10] Sunspots are relatively small, dark areas that appear on the sun's surface. In reality, they are great magnetic storms that swirl through the hot, gaseous *photosphere* of the sun. Sunspots have been closely observed—and coun-ted—by astronomers since about 1610, roughly the time the telescope was introduced.

According to the theory of the solar researchers, a greater number of sunspots generally appear when the sun is more "ac-tive," that is, when it is putting out greater radiation. The stronger the radiation that reaches the earth, the warmer the earth becomes; weaker radiation leads to a cooler earth.

Dr. Stephen H. Schneider of the National Center for Atmospheric Research (NCAR), and Clifford Mass, University of Washington, decided to test the possible relationship between sunspot numbers and the radiative output of the sun. They developed a mathematical formula from which they could calculate the earth's surface temperature, using not only sunspot numbers but also the effect of volcanic dust veils. A volcanic *dust veil index* had been compiled and tabulated by Professor Lamb. He used both historical accounts of volcanic eruptions and direct measurements of volcanic dust in the atmosphere to estimate the extent to which solar radiation might have been reduced, for each year back to 1600.

The calculations of Schneider and Mass (see figure 3) show that a major drop in global temperature should have taken place around the year 1620, with the chill hanging on until the early 1700s. History verifies that taking place, of course—the "Little Ice Age."

The Schneider and Mass calculations indicate that another drop in the earth's temperature should have occurred just after 1800. That calculated cooling, not so marked as the previous one, coincides with the infamous "Year Without a Summer." So, the assumption, or theory, that sunspot numbers are indicative of the strength of solar radiation appears to be a valid one.

Interestingly enough, the two significant cold periods modeled by Schneider and Mass are initiated about 180 years apart. That supports the idea that the 180-year cooling cycle, as evidenced in the ice cores studied by Dansgaard, might be caused by a variation in the solar constant and concomitantly the amount of radiative energy reaching the earth.

Studies of a radioactive carbon isotope found in trees also suggest quite strongly—as do the Schneider and Mass calculations—that there is an important relationship between sun-

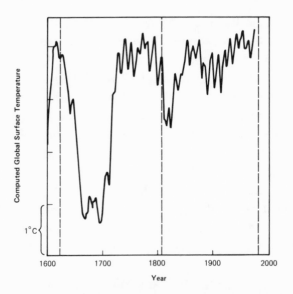

3. Global temperatures as calculated by Schneider and Mass. Their mathematics show that significant global cooling should have occurred around 1620 and again near 1800, or about 180 years apart as indicated by the broken vertical lines. History verifies that those cooling trends did, in fact, take place. (Dr. Stephen H. Schneider and Clifford Mass, *Science*, vol. 190, pp. 741–746, fig. 2, November 21, 1975. Copyright 1975 by the American Association for the Advancement of Science.)

spot numbers and solar variability. Carbon 14 is formed in the earth's atmosphere by cosmic rays, and is assimilated into trees in the form of carbon dioxide (CO_2). The amount of cosmic ray activity reaching the earth is controlled by solar magnetic phenomena. When the sun is more active, its magnetic field extends—remember, sunspots are magnetic disturbances—and shields the earth from some of the cosmic rays heading this way from space. Fewer cosmic rays reach the earth's atmosphere and less carbon 14 is formed.

When the sun is not so active, its magnetic field weakens. More cosmic rays strike the earth's atmosphere, and the production of atmospheric carbon 14 increases. Thus, John A. Eddy, an astrophysicist at NCAR, reasoned that a prolonged period of inactivity on the sun should be manifested in a higher carbon-14 content in trees.[11] For the period 1645–1715, near the nadir of the "Little Ice Age"—when the sun was relatively inactive and there was almost no sunspot activity—analyses do indeed show a 20-percent increase in the carbon 14 measured in the annual growth rings of trees!

The carbon-14 records show another period of solar quiet began about 180 years before the "Little Ice Age" and lasted from around 1460 to 1550. This correlates with the cool period on earth suggested by the ice cores to have begun around 1440 and which may have led to the disappearance of the colonies on Greenland.

Work by Paul E. Damon, a University of Arizona scientist, indicates that regular, periodic increases in the carbon-14 content of trees have been commonplace well back beyond the two most recent periods of solar quiet. Damon found a significant 182-year cycle extending back 2,000 years from the present.[12]

Thus, there seems to be a good deal of evidence that the cooling cycle of about 180 years, as displayed in the Greenland ice samples, may be related to a periodic fluctuation in the energy output of the sun, which in turn can be related to varying sunspot numbers.

With that in mind, and since we are nearing the next 180-year point (the early 1980s), it is interesting to compare recent solar activity with that of 180 years ago. Sunspot activity progresses through an irregular cycle of roughly 11 years during which the number of spots on the sun moves from a minimum to a maximum and back. The number of actual sunspots may be as low as 0 or as high as 200 on an average annual basis. And the cycle may vary in length from 10 to 15 years, around its average of 11.

The cycle of solar activity that began in 1964 and apparently ended in 1976[13] was defined as cycle 20; the corresponding cycle 180 years earlier was cycle 4. Cycle 20 was the longest of this century. As you can see from figure 4, the correlation between cycles

4. Comparison of sunspot cycle 20 with sunspot cycle 4. Cycle 20 was the longest of this century. Cycle 4 was also exceptionally long and preceded the last significant global cooling. (Dr. Patrick S. McIntosh, National Oceanic and Atmospheric Administration, Space Environment Laboratory.)

20 and 4 is remarkable.[14] Cycle 4, also exceptionally long (nearly 15 years), preceded the last significant global cooling—that cooling calculated by Schneider and Mass, evidenced in the ice cores, and manifested by the cold summer of 1816, among other events. Cycle 4 was followed by two sunspot cycles (5 and 6) that had unusually low numbers of sunspots during sunspot maxima (see figure 6). Low sunspot numbers: diminished solar constant: cooler temperatures on earth. The coincident patterns of cycles 20 and 4 would seem to suggest that cycles 21 and 22 may behave very much like cycles 5 and 6.[15] The implications for our immediate climatic future are clear.

A Sunspot-Climate Cycle Hypothesis

Dr. Hurd C. Willett, professor emeritus at the Massachusetts Institute of Technology, has been studying the correlation between solar activity and the earth's weather for over 30 years. He sees a definite relationship between sunspot cycles and the climate regimes that affect us. On a global scale these climate regimes are determined by the configuration of the earth's wind patterns. The wind pattern that is most important, at least for the purposes of this discussion, is the one commonly known as the *jet stream*. Dr. Willett prefers the term *zonal westerlies of middle latitudes* rather than *jet stream*. This is because the jet stream is actually only a small part of the zonal westerlies (or just westerlies), the name given an endless band of upper atmospheric westerly winds circling the hemisphere at mid-latitudes (30°–65°N).

The jet stream is the axis of the strongest westerlies. It usually howls along at around thirty thousand feet and may reach speeds up to several hundred miles per hour. In general, the westerlies are the reason a Boeing 747 can fly from Los Angeles to New York faster than it can from New York to Los Angeles.

The westerlies are part of what meteorologists call the *general circulation* of the atmosphere, which is what Dr. Willett is talking about when he says, "Climatic fluctuations are best expressed in terms of fluctuations of the general circulation pattern. . . ."[16] What he means is that climatic variations, or changes, occur as the pattern of overall atmospheric circulation changes. Changes in the configuration of the westerlies are reflective of changes in the overall pattern.

In reality, the westerlies do not blow from west to east all of the time. More typically they display an undulating pattern, sweeping southward over one part of the hemisphere, and back northward somewhere else. Occasionally they may form a complete loop. The undulations and loops constantly change positions, and it is these shifts in the configuration of the westerlies that produce changes in our weather regimes.

Where the westerlies shift or loop northward, they produce relatively warm, dry weather. Where they bend southward, generally cooler, wetter conditions result. Figure 5 presents a schematic representation of the westerlies and how they effect North American climate.

Dr. Willett has pointed out the existence of a long-term sunspot cycle in addition to the basic 11-year cycle, and he has related

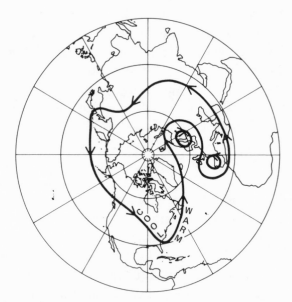

5. Undulating pattern of the westerlies; looping (or blocking) patterns are indicated over western Europe; the pattern shown over North America would produce warm, dry weather in the western United States, cool, wet conditions in the East.

various phases of that cycle to changes in the pattern of the westerlies. Changes in the configuration of the westerlies, which last over a period of decades, bring about changes in climate. Willett thinks that in this manner the long-term sunspot cycle influences climatic trends on earth.

The long-term cycles alternate between 80 and 100 years in length. You will note that two of the long-term cycles equal 180 years, precisely the length of the cycles that I discussed in the previous section. Not only is the length the same, so is the phasing, with the last 80-plus-100-year couplet having begun about 1800 and the next one due to start about 1980. Figure 6 indicates the periods of the long-term cycle (80 and 100 years) as well as the number of sunspots for each 11-year cycle within the long cycle.

Willett contends that the amplitude—as defined by the sunspot number—of each 11-year cycle repeats in virtually the same sequence at the beginning of each long-term cycle and that climatic patterns on earth follow a similar long-term repetition pattern.

Specifically, Willett suggests that during the initial 2 or 3 decades of each long-term solar cycle the westerlies shift farther south, thus producing maximum coldness at all latitudes, except in equatorial and subtropical regions. If you will look at figure 6 again, you will see that sunspot numbers are at a minimum near the beginning of each long-term cycle. So, the southward shift of the westerlies at that point would be consistent with the theory of reduced sunspot numbers being related to a diminished solar constant, which in turn produces, on a global scale, cooler weather.

When you examine figure 6 and consider the other 180-year cycles that I have discussed, you might conclude that the maximum effect—cooling—of any reduction in the solar constant would be felt only at the beginning of the 80-year cycle, with

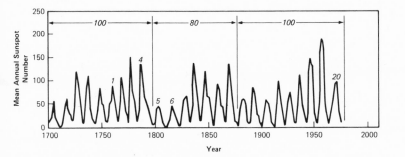

6. Annual smoothed sunspot numbers indicating the number of sunspots for each 11-year cycle. Periods of long-term solar cycles are shown across the top. Cycle numbers are indicated by digits near the peaks of various cycles. Professor Willett contends that the amplitude of each 11-year cycle repeats in virtually the same sequence at the start of each long-term cycle, and that climatic patterns on earth follow a similar long-term repetition pattern.

relatively minor effects apparent at the onset of the 100-year cycle. That, indeed, might be so. On a qualitative basis, though, the effect would be the same at the start of either cycle: The westerlies move south and we get colder.

The westerlies reached their maximum northward extent in the early 1930s, producing a global warming trend. This is reflected by the fact that a greater number of state records for high temperatures and dryness were set during the 1930s than during any other decade since the 1870s (see table 1).

After the early 1930s the westerlies began to display a greater frequency of looping (or blocking) patterns, and we went into what Willett calls a period of *climatic stress*, which lasted through the 1950s. When a blocking pattern is established, some regions become persistently hot and dry, others cold and/or wet. Table 1 shows that not only were the greatest number of state records for high temperatures set in the 1930s, so were the greatest number of state record lows. The largest number of states reporting record wetness was in the 1950s; so was the second greatest amount observing record dryness—climatic stress.

Another consequence of the westerlies shifting to a higher than normal latitude during the 1930s was that hurricane frequency reached a peak. Warm tropical oceans and an absence of westerly winds are favorable for hurricane development. Twenty-one tropical storms and hurricanes popped up in the Atlantic, Caribbean, and Gulf of Mexico in 1933, and seventeen in 1936. The contemporary (1946–1975) average is nine. The climatic stress period that followed the early 1930s brought several destructive hurricanes into New England in the 1950s, while in Florida the number dwindled to a record low level.[17] It is apparent, then, that solar cycles affect not only temperature and precipitation regimes but also the character of hurricane seasons.

In work that seems to corroborate Dr. Willett's theory Professor Lamb has noticed a correlation between sunspot activity and the frequency of westerly winds in England.[18] Lamb points out that

TABLE 1

State* Record Maximum Temperature Decade	*Number of state records*	*State* Record Minimum Temperature* Decade	*Number of state records*
1870–1879	0	1870–1879	0
1880–1889	1	1880–1889	1
1890–1899	2	1890–1899	7
1900–1909	2	1900–1909	6
1910–1919	4	1910–1919	5
1920–1929	3	1920–1929	3
1930–1939	26	1930–1939	12
1940–1949	1	1940–1949	5
1950–1959	7	1950–1959	5
1960–1969	1	1960–1969	5
1970–1976	2	1970–1976	1

State*Record Maximum Annual Rainfall Decade	*Number of state records*	*State* Record Minimum Annual Rainfall* Decade	*Number of state records*
1870–1879	2	1870–1879	0
1880–1889	5	1880–1889	0
1890–1899	2	1890–1899	0
1900–1909	3	1900–1909	0
1910–1919	1	1910–1919	1
1920–1929	1	1920–1929	1
1930–1939	3	1930–1939	18
1940–1949	8	1940–1949	6
1950–1959	11	1950–1959	11
1960–1969	7	1960–1969	11
1970–1976	1	1970–1976	0

Other records in 1840, 1845, 1851, 1853, 1869

Other record in 1826.

*Includes District of Columbia but not Alaska and Hawaii.

"remarkably high frequencies" of westerlies seem to precede major sunspot maxima (such as occurred in 1960 and 1780) by about 30 or 40 years. He goes on to say the westerlies then drop off sharply in frequency—that is, shift to a lower latitude—during the next several decades, reaching a minimum from which they do not recover until approximately 150 years later. Thus again, there is the suggestion of a 180-year period (30 plus 150 years).

The 20-Year Cycle

Another cycle of importance, at least over eastern North America,[19] and perhaps the rest of the continent, too,[20] is a 20-year cycle in wintertime temperatures. That cycle is most prominent in January, and is best reflected in 10-year *running means*. These are 10-year averages shifted 1 year ahead for each calculation. Thus, the 10-year running mean for 1977 would be for the period 1968 through 1977, and for 1978 would be for the period 1969 through 1978. Figure 7 shows the 20-year cycle for Boston, Massachusetts. The running means indicate the cycle consists of a 10-year stretch of relatively warm Januarys, followed by a 10-year period with relatively cold Januarys. Because 10-year running means are used, a smoothing effect occurs and short-term variations are filtered out, highlighting the long-term effects. Thus, it is not unlikely that a cold January could appear in a warm period, and vice versa. The fact that the same cycle occurs at the same time (at least from 1890 to about 1960) over a large geographical region is displayed in figure 8, which graphs the January cycle for twelve stations in eastern North America.

The 20-year cycle also exists in the West, the Pacific Northwest in particular.[21] The cycle there is not quite so striking and well defined as it is in the East, but it does show up, occurring in almost direct opposition to the phase of the cycle in effect in the East. That is, the cold phase in eastern North America coincides with the warm phase in the West. And the eastern warm phase

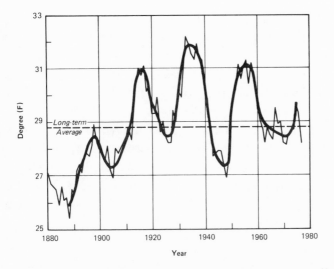

7. The Boston 10-year running mean January average temperatures. The running means indicate the presence of a 20-year cycle. That cycle consists of a 10-year stretch of relatively warm Januarys, followed by a 10-year period with relatively cold Januarys.

happens when it is relatively cold in the West.

That the phase of the cycle is opposite on opposite sides of the continent is not a mystery. It can be explained by the positioning of the undulations in the westerlies. When it is cold in the West and warm in the East, the westerlies show a tendency to blow from the northwest over the West, then pivot around to blow from the southwest over the East. The northwesterly undulation in the western United States drives cold, Canadian air masses into that part of the country. In the East, the southwesterly winds sweep mild air northward from the Gulf Coast states. When the cycle changes phase, so does the configuration of the westerlies. Such a change sees a southwest flow develop over the West while a northwesterly tendency manifests itself in the East. Result: warm West, cold East.

The particular pattern of the westerlies that produces each

phase of the 20-year cycle doesn't stay locked in year after year. But over a 10-year period it would be a frequently repeated or favored configuration.

The cause of the 20-year oscillation is not known. But there is strong evidence that it may be produced by sea surface temperature anomalies—that is, areas of unusually warm and cold water—in the Pacific Ocean.[22] Some scientists have tried to relate the 20-year cycle to the double sunspot cycle, two 11-year cycles. Such a relationship holds up very convincingly in this century but seems to falter some in the 1800s.[23] It's not necessary to argue that point here. What's important is the *fact* of the 20-year cycle and how it appears to complement the 180-year cycle.

A Conjunction of Cycles

As I pointed out earlier, forward projection of the 180-year cycle indicates the earth is on the brink of a significant cooling. Willett's work with solar-climatic relationships and the similarity of sunspot cycle 20 to sunspot cycle 4 (figure 4) seem to indicate the same thing: The next couple of decades are going to be generally colder. Not a new ice age. But enough to make winters snowier and/or fuel bills higher in many areas.

If the 20-year cycle is considered, the implication is that the cooling will initially be felt much more strongly in the eastern United States than in the West. For instance, the projection of the cycle for Boston (figure 7) suggests a period of colder winters (the 180-year cycle aside) beginning in the early 1980s. However, you will notice from figure 7 that the 20-year cycle displays some rather strange behavior after about 1960. The well-defined phases of relatively warm and cold Januarys disappear, and an irregular oscillation sets in. The same type of thing shows up in figure 8: The cycles at various locations suddenly drop out of phase after about 1960.

Thus, the forward projection of the 20-year cycle at this point

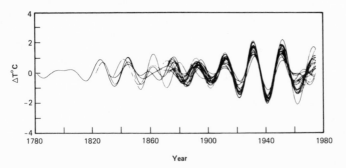

Year

8. The 20-year cycle as depicted in January mean temperatures for twelve stations in eastern North America. It is apparent that the same cycle occurs at the same time over a large geographical region. (Dr. Stephen A. Mock, *Nature*, June 10, 1976.)

might seem a rather foolish undertaking. Indeed it might be. But it's perhaps significant that the last time the cycles drifted out of phase was around 1860, 20 years prior to the end of the last long-term solar cycle. At that same time, the 10-year running means of January temperature at New Haven, Connecticut, where continuous records date back to 1781, began precisely the same type of irregular oscillation that is evident in the Boston graph now. From 1810 to 1860 the 20-year cycle had been clearly defined (see figure 8, the longest record is that of New Haven). After 1880, which marked the advent of the current long-term solar cycle, the 20-year cycles phased in again over a large geographical region, and the 10-year running means once more exhibited their smooth 20-year oscillation.

All of this may be important, of course, because 1960 was 20 years prior to the predicted end of the current long solar cycle. Thus, we might expect that the irregular oscillation will soon stop and that the cycles will drop back in phase around 1980, the start of the next long-term solar cycle. The further implication is that over eastern North America the cooling phase of the 20-year cycle will be superimposed on the 180-year cycle.

Before we leave the 20-year cycle, one more item is worthy of

TABLE 2   Summary of Climatic Clues

Clue	Implied effects	When effects should first be apparent
180-year ice core cycle	Global cooling	1980s
Long-term sunspot cycle	Low-latitude westerlies leading to global cooling	1980s
Similarity of sunspot cycles 4 and 20	Diminished "solar constant" during cycles 21 and 22 leading to global cooling (may be tied to long-term sunspot cycle)	Early 1980s
20-year cycle in January means	Predominately colder Januarys in eastern U. S. (would augment 180-year cooling cycle)	Early 1980s

mention. The cooling swing of the 20-year cycle did, in fact, coincide with the onset of the influence of the last 180-year cycle (around 1800): The running means at New Haven indicated a *20-year*—not 10-year—period of cold Januarys from 1803 to 1833.

Summing up then, the clues point toward noticeably colder winters (Januarys) in the East for about a decade, or perhaps two, beginning in the early 1980s. The effect in the West would probably be somewhat less apparent, with the maximum cooling perhaps delayed for 10 or 20 years. Table 2 summarizes the clues and their implied effects.

A Look at the Past

On a broader scale the compelling, albeit circumstantial, evidence indicating that we are about to experience a significant global cooling trend—the 180-year cycle—suggests that our weather of the immediate future will be more like that of the early 1800s than that of the recent past. The National Weather Service (NWS) defines current climatological normals as being the averages from the period 1941 to 1970. (The NWS always uses the 30-year period ending prior to the current decade in order to calculate normals.) To get a better picture of the climate in store for us over the next couple of decades, we should probably be looking at averages from 1800 through 1829. Unfortunately, comprehensive and continuous records for most locations weren't kept until after about 1870. But the search for parallels with the past is not hopeless.

You will remember Dr. Willett indicated that the westerlies drop to lower than normal latitudes during the first several decades of each long-term solar cycle. Thus, if we look at the weather records from the initial decades of the *current* long-term cycle (which began about 1880; look at figure 6 again), we can get a good idea of the type of climate patterns to be expected with the onset of the *next* long-term cycle. The beginning of the next long solar cycle coincides with the 180-year cooling cycle, of course. Dr. Willett observes that during the 1880–1910 period, the westerlies were indeed at prevailingly low latitudes.[16] In response to that pattern, Northern Hemisphere temperatures were at their coolest levels in the past 100 years (figure 9).

You might argue, and maybe with good justification, that if I am going to claim that the next couple of decades will be climatically similar to the early 1800s, then using data from the late 1800s would be a little conservative. That is, by doing that, I would be underestimating any cooling. On the other hand, I would point out that the current global temperature average is closer to that of just before 1880 than to that of the late 1700s.[24] In other words, the climate regimes around 1900 began with temperatures more similar to what we have now than did the regimes of the early 1800s.

Others besides myself have suggested that we ought to be

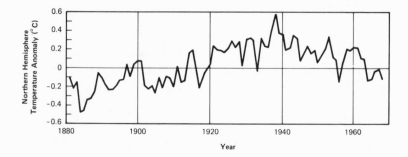

9. Northern hemisphere temperature trends since 1880. In response to prevailing low-latitude westerlies during the period 1880–1910, northern hemisphere temperatures were at their coolest levels in the past hundred years. (After Dr. M. I. Budyko, *Tellus*, vol. XXI (1969), 5, p. 612, fig. 1.)

analyzing weather records from at least the late 1800s to get an idea of what lies in our immediate future. For instance, Professor Lamb has indicated contemporary climate normals may no longer be valid and says, "The climatic statistics of times before 1895 may . . . be more relevant to the present day and for some decades ahead."[18]

J. Murray Mitchell, a senior research climatologist for the government, also feels that we should be examining historical weather records. He emphasizes his point by saying, "We've gotten burned time and time again using just modern weather records."[25]

Historical Records

Beginning in 1870 a large-scale federal weather service was authorized by Congress. The service was originally run by the U.S. Army's Signal Service. In the 1890s the function was transferred to the newly created Weather Bureau under the Department of Agriculture.[26] Generally speaking, the climatological observations around the turn of the century were complete and made quite carefully. One state climatologist whom I questioned on the reliability of the records suggested that the observations of 80 and 100 years ago were perhaps taken even more conscientiously than those of today.

Still, there are certain problems that crop up when comparing averages of 100 years ago with contemporary averages from the same location. The most important of those problems stems from urbanization, the growth of large cities around weather observing sites. Big metropolitan areas act as "heat islands." Other things being equal, the "heat island" effect produces somewhat warmer temperatures in cities than in surrounding countrysides. Just how much this warming has influenced average temperatures is open to question.

It is difficult, if not impossible, to separate the "heat island" effect from natural temperature trends, and trends induced by changes in thermometer exposure, height, and location within the city. In fact, many of the observing stations have been moved in recent years from cities to suburban airports. Therefore, in some records, a reverse "heat island" effect may be manifest.

One measure of how "real" temperature trends are would be to see if similar trends are exhibited by stations situated in different environments within the same geographical region. For instance, New York City (Central Park), Boston (city, then airport located on Boston Harbor), and Blue Hill Observatory (suburban Boston) all display temperature trends of similar direction and magnitude over the past century.

Boston, for example, has a contemporary average annual temperature that is about 2 degrees warmer than was the annual mean at the turn of the century. That a good part of that difference is natural is suggested by the fact that a greater percentage of the wintertime precipitation fell as snow in the period 1880–1909 than in the period 1941–1970 (see chapter 11). The point of that argument is that if the warming had been due strictly to urbanization, a local effect, then the percentage of precipitation falling as snow should not have changed: The determination of rain versus snow is governed by large-scale atmospheric processes. The fact that the percentage has changed indicates that Boston's temperature regime has been altered by something more than a local influence.

All effects considered, it would seem as if we can accept the differences in climatological averages from different periods at the same location at pretty much face value.

The Future

The following chapters take the hypothesis that the weather of the next few decades will be more similar to that of the late 1800s and very early 1900s than to that of the recent past, and change it into numbers and descriptions for each region of the United States, giving answers to the question, "What will the weather be like where I live?"

Get ready for some substantial changes.

Notes

1. Eddy, J. "The Case of the Missing Sunspots," *Scientific American* (May 1977); and "The Maunder Minimum," *Science* 192 (1976): 1189.
2. Pruitt, L. "Ice Cores: Clues to Past Climates," *Science News* 98 (1970): 369.
3. Broecker, W. "Climatic Change: Are We on the Brink of a Pronounced Global Warming?" *Science* 189 (1975): 460.
4. Schultz, G. *Ice Age Lost* (New York: Anchor Press/Doubleday 1974), p. 182.
5. Flohn, H. *Climate and Weather* (New York: McGraw Hill 1969), p. 211.
6. Landsberg, H. and Albert, J. "The Summer of 1816 and Volcanism," *Weatherwise* 27 (1974): 63.
7. Lamb, H. speaking in a segment of *The Weather Machine*, a 2-hour educational television program first aired on BBC2 in 1974 and later over U.S. television.
8. Lamb, H. "On the Nature of Certain Climatic Epochs Which Differed From the Modern," *Proceedings of the WMO/UNESCO Rome (1961) Symposium on Climate Changes (Arid Zone XX)*, UNESCO (1963), p. 125.

9. Hammond, A. "Solar Variability: Is the Sun an Inconstant Star?" *Science* 191 (1976): 1159.

10. Schneider, S. and Mass, C. "Volcanic Dust, Sunspots, and Temperature Trends," *Science* 190 (1975): 744.

11. Eddy, op. cit.

12. Damon, P. "Solar Activity Induced Variations of Energetic Particles," University of Arizona, Department of Geosciences Contribution No. 714 (1976).

13. Personal correspondence with Patrick S. McIntosh, Space Scientist, NOAA Space Environment Services Center, January 1978.

14. Purrett, L. "Autumn on the Sun," *NOAA*, July 1976, p. 22.

15. Because in early 1978 cycle 21 was showing a definite departure from the pattern of cycle 5, the official view of the U.S. Space Environmental Services Center was that there was only a 5 percent probability of cycle 21 being as low as cycle 5. However, H. Prescott Sleeper, Jr., of Kentron International has developed a solar activity model that indicates a repetition of the magnitude of cycles 5 and 6 will likely occur. Sleeper feels that the new cycle begun in 1976 may well be abortive, and that another sunspot minimum will be reached before the end of 1979.

    Sleeper, H. P. "Planetary Resonances, Bi-stable Oscillation Modes, and Solar Activity Cycles," NASA Contractor Report CR-2035 (1972); and personal correspondence with H. P. Sleeper, January 1978.

16. Willett, H. "Do Recent Climatic Fluctuations Portend an Imminent Ice Age?" *Geofisica Internacional* 14 (1974): 265.

17. Dunn, G. and Miller, B. *Atlantic Hurricanes* (Baton Rouge: Louisiana State University Press, 1964), pp. 46–49.

18. Lamb, H. "Climate in the 1960's," *Geographical Journal* 132 (1966): 183.

19. Mock, S. "The 20-yr Oscillation in Eastern North American Temperature Records," *Nature* 261 (1976): 484.

20. Currie, R. "Solar Cycle Signal in Surface Air Temperature," *Journal of Geophysical Research* 79 (1974): 5657.

21. Pica, G. "Weather Cycle," *Eugene Register-Guard*, 29 December 1976, p. 1D.

22. Namias, J. *Short Period Climatic Variations, Collected Works of J. Namias, 1934–1974* (San Diego: University of California, San Diego, 1975).

23. Bernard, H. "A Mini Ice Age Could Begin in Decade," *Globe*, 2 November 1975, p. 6.

24. Budyko, M. "The Effect of Solar Radiation Variations on the Climate of the Earth," *Tellus* 21 (1969): 611.

25. Ognibene, P. "Is Bad Weather Ahead? Plan On It," *Parade*, 30 October 1977, quoting J. Murray Mitchell, Jr.

26. Henry, A. *Bulletin Q, Climatology of the United States*, U.S. Government Printing Office, 1906, p. 6.

CHAPTER 3

The Coming Climate: What It Means for Droughts, Hurricanes, and Tornadoes

IN ADDITION TO changes in rainfall, snowfall, and temperature, some shifts in the patterns of the more dramatic weather events—droughts, hurricanes, and tornadoes—can be expected in the coming decades. Notice that I used the word *patterns*. I do not want to imply any change in the *intensity* or *severity* of droughts, hurricanes, or tornadoes. What I do want to suggest is that the area of maximum threat for these various weather headline makers will shift. Such shifts can be postulated from the change in the westerlies that I talked about in the preceding chapter.

But even more important than postulates, or theoretical considerations, are weather and climate records of the time period 1880–1909, to which I expect the weather of the coming decades to be analogous. If the information in these records supports the postulates, so much the better. In some cases the historical records extend back beyond 1880. For instance, a number of researchers, by examining the growth histories of certain types of trees, have reconstructed western United States drought patterns that date

back to 1700. And David M. Ludlum, long-time editor of
Weatherwise and America's preeminent weather historian, has
documented a chronology of American hurricanes that begins
with Christopher Columbus's first voyage to the New World in
1492.[1] The value of such long-term records is that they provide information from the time of the last bottoming out of the 180-year
cycle: the early 1800s.

Droughts

On a broad time scale, the occurrences of major droughts in the
United States west of the Mississippi River are relatively easy to
predict. This is because they seem to follow a fairly regular cycle
of 20 to 22 years, and match up very closely with the double sunspot cycle, two 11-year cycles. As figure 10 shows, major droughts
in the western United States have occurred in conjunction with
every other sunspot minimum back to 1700. In this century the
drought that led to the Dust Bowl days of the 1930s is part and
parcel of American history; severe drought stalked the Great
Plains again in the mid-1950s; and the California and Rocky
Mountain droughts of the mid-1970s were etched indelibly on our
memories by the medium of television.

By extrapolating the double sunspot cycle, meteorologists can
say with some certainty—but not absolute certainty, for cause and
effect is still a mystery—that the next major drought in the West,
assuming that that of the 1970s has departed, will occur around
the year 2000. What cannot be forecast is *exactly* when the
drought will occur or precisely where it will occur. Although I
won't venture a guess as to the starting date of the next important
western drought, nor how long it might last, I will, with the help
of Dr. Willett and a number of dendrohydrologists (scientists who
study climate records as revealed in tree rings*), suggest what

*This is an entirely different function than examining tree rings for carbon-14
content to deduce the amount of solar activity that occurred during a particular
period (chapter 2).

region is likely to be most severely affected by it.

Dendrohydrologists are important in this undertaking because they are the people who reconstruct drought patterns back through history. They do this by examining annual growth rings of trees. A tree usually grows somewhat more slowly during an extended dry period, thus producing narrower than normal growth rings. The widths of these narrower than normal rings are compared with a width that would be expected in a normal growth year, and numerical ratios, or *tree ring indices*, are calculated. A series of tree ring indices averaged from a number of trees in one location can thus give a history of drought in that location. Although other phenomena besides drought, such as fire or insect damage, can alter tree growth, it is by and large climatic factors that are the most important.[2]

One contemporary measure of drought is called the *Palmer drought severity index (PDSI)*, which is computed from temperature and precipitation data as recorded by instruments. By mathematically correlating tree ring indices with the PDSI,

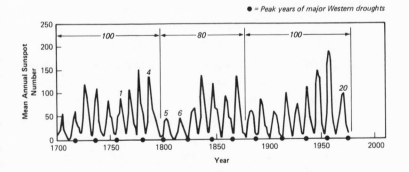

10. Annual smoothed sunspot numbers with the peak years of major western droughts indicated by dots. These major droughts have occurred in conjunction with every other sunspot minimum back to 1700. The next severe western drought—assuming that that of the 1970s has departed—will probably take place around the year 2000.

researchers have been able to verify that tree ring growth patterns can indeed accurately reflect drought conditions lasting for at least 2 years. Further, on the basis of the correlations, they have been able to use the PDSI to describe historical droughts. That is, they can assign an index number to a drought, say one which occurred in the early 1800s, for which there are no temperature and precipitation measurements available. Interestingly enough, they arrived at the conclusion that the driest single year since 1700 in the western United States was 1934.[3]

Dr. Willett suspects that over the next several decades there will be no major prolonged drought in the United States, except perhaps in the area from southern California eastward through Arizona and New Mexico into western Texas. He goes on to speculate that a severe drought across the Canadian plains may occur concomitantly with the one in the southwestern United States.[4]

But just how does a shift in the westerlies make one area more vulnerable to drought than another? Remember, we're talking about a southward shift of the westerlies over the next 20 or 30 years. Such a shift would move the primary storm tracks and their attendant precipitation patterns a little southward, too. Thus, adequate rain and snow would fall over most of the northern and central United States. But those areas might benefit at the expense of our neighbor to the north. As the storm paths vacate Canada more frequently in favor of the United States, the Canadian prairies would be dominated by cold and relatively dry polar air masses. Less and less precipitation would fall, and drought would build.

But the problem in our own Southwest would not be caused by the dominance of polar air. Another factor would be working against such states as Arizona and New Mexico. An upper-air high-pressure area that resides on a semipermanent basis over the Interior West would be shunted southward as the westerlies move southward. This high pressure is always found south of the main band of westerlies and produces generally clear skies and dry weather. In a high-pressure area, or *anticyclone*, air descends, sliding down off the mountain of air producing the high pressure. Sinking air heats up and dries out under the influence of its own compression. Thus, clear skies and dry weather. With the westerlies moving southward, the effects of the western upper-air high would be concentrated on the Southwest.

That is the theory. But what does climatic history tell us? The period that would give us the best clue would be the early 1800s, as we discussed in chapter 2. Unfortunately, there were not too many rain gauges in western North America at that time. That's why dendrohydrologists are so important. According to the double sunspot cycle, the major drought period in the early 1800s should have been in the 1820s (see figure 10). In fact, that was when it did occur, and it was centered in New Mexico and Arizona! (See figure 11.) At the same time, tree ring analysis indicates that precipitation was above normal in the northern Rockies and northern Great Plains. Tree rings also tell us that the first 2 decades of the nineteenth century were wetter than normal in the Pacific Northwest.[2,3] Of course, the basic thesis of this book is that the period 1880–1909 should give just as good a picture of our coming climate as the early 1800s. Dendrohydrologists tell us that in the years between 1880 and 1909 drought was most severe and persistent in southern California, Arizona, and Nevada. Occasional incursions of drought reached into the northern Rockies and eastern Great Plains, but it was the Southwest that again suffered the most.[2,3] (See figure 12.)

Dendrohydrology studies also suggest that in the 1640s, going back through two 180-year cycles, drought areas shifted even farther south, into Mexico.[2]

Thus, theory and observation seem to agree: It should be Arizona, along with southern California, New Mexico, and Mexico, that had best be prepared for major drought as the peak of the next 20- or 22-year cycle approaches.

The U.S. Water Resources Council, an independent federal agency, estimates that all of the western United States will have a severe water shortage by the year 2000.[5] It is obvious that a prolonged drought about that time would only exacerbate the problem. Arizona, in particular, has been identified as a critical area. It is the fastest-growing state in the nation and is almost entirely dependent upon underground water for both irrigation and municipal use. The water comes from an ancient glacial lake underlying the state. The water in that lake has been built up over eons but is now being pumped out three times faster than nature, through rainfall and snowmelt runoff, can replenish it. Because of Arizona's population boom, the pumping rate will probably continue to increase, and the gap between use and replenishment widen.

Given that fact and given a major drought in the Southwest, economic growth in Arizona could be brought to a virtual halt. Such a threat would seem to emphasize the importance of completion of the Central Arizona Project (CAP). CAP is a two-

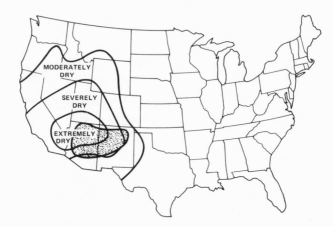

11. This is a "snapshot" of drought conditions in July 1822. It is based on a dendrohydrological reconstruction using the Palmer drought severity index. The shaded area represents the primary drought region from 1816 through 1825. (After Drs. Charles W. Stockton, David M. Meko, and Harold C. Fritts.)

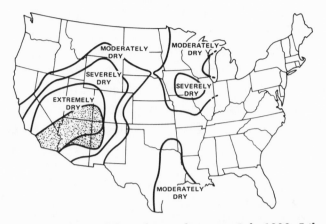

12. This is a "snapshot" of drought conditions in July 1902. Like figure 11, it is based on a dendrohydrological reconstruction using the Palmer drought severity index. 1902 was one of the years in which the drought reached into the Rockies and eastern Great Plains. The shaded area represents the primary drought region from 1896 through 1905. (After Drs. Charles W. Stockton, David M. Meko, and Harold C. Fritts.)

billion-dollar Bureau of Reclamation project designed primarily to deliver Colorado River water to the burgeoning cities of Phoenix and Tucson.

If CAP is not completed, water normally used by Arizona agriculture—agriculture consumes about 85 percent of the annual water supplies in the West—might have to be transferred to the cities. Such a diversion could be disastrous to southwestern agribusiness. The economic effects would be felt nationwide, of course, particularly in rising prices of such items as beef, lettuce, and cotton products.

As of early 1978 President Carter had made delivery of CAP water conditional upon the implementation of a strict water management program in Arizona. The adoption of such a plan would seem mandatory in the face of Arizona's water problems, not to mention the potential for a severe drought.

Even if the CAP is completed, much of its water would come at the expense of southern California, Los Angeles in particular. Southern California must rely almost exclusively on "imported" water from the Colorado River and from precipitation runoff in the Sierra Nevada mountains of northern California. Los Angeles is currently using Arizona's share of Colorado River water.

The shift of that water back to Arizona upon completion of CAP would leave Los Angeles with a significant water shortage. And because any drought menacing the Southwest would certainly be a threat to southern California (see figure 12), that area's search for other water sources may take on an acute urgency. The alternatives are not promising.

Water from the Columbia or Snake River basin could be shunted to southern California from the Pacific Northwest. But a moratorium on any such studies, let alone projects, was mandated through 1978 by the Colorado River Basin Project Act of 1968, which authorized CAP.

Cloud seeding might increase precipitation in certain areas, but legal problems and so-far-doubtful results make that alternative less than attractive. After a 5-year cloud-seeding program in the San Juan Mountains of the upper Colorado River basin, the Bureau of Reclamation reported that there was "no significant added precipitation."[6]

The desalination of sea water has also been considered as a method of increasing water supplies. But skyrocketing energy costs have resulted in substantial increases in the costs of desalinated water. Desalination on a large scale does not seem to hold an economical answer to water augmentation.

After examining these and other alternatives, a 1977 National Academy of Sciences study reported that it appears "no significant . . . augmentation is available in the immediate future."

Thus, it seems the solution to Arizona's water supply problems may be the beginning of southern California's. That, of course, assumes CAP really is the ultimate solution to Arizona's dilemma.

Quite probably it is not. By the turn of the century, it is questionable whether there will be enough water in the Colorado River every year to satisfy all of the proposed energy and agribusiness development in the United States, plus Indian and Mexican claims.[6]

So, in the teeth of a severe drought around the year 2000 —especially one that might extend into Utah, Colorado, and Wyoming (see figure 12), from whence comes most of the water in the Colorado River—there may be little that can be done to avert serious water supply problems in the southwestern United States. Acute drought, as the new century starts, should be centered in Arizona and New Mexico. But that prediction doesn't leave much margin for error. A northward shift of the drought region by a few hundred miles, into the important watershed lands of the upper Colorado River basin, would raise havoc with water allocations in one of the fastest-growing sections of the nation. It is a real worry.

Hurricanes

The prediction that the westerlies will slip southward, to slightly lower latitudes, over the next few decades has important implications for the hurricane belt. The most important implication is that relatively fewer of the more severe or damaging hurricanes should strike areas north of Cape Hatteras, North Carolina, while regions to the south may be visited somewhat more often.

Quite simply, the theory here is that the westerlies, the large-scale wind currents that steer most storms in the mid-latitudes, will more often than not be in position to force major hurricanes to follow the classic *recurvature path*: moving toward the U.S. coastline from the south or southeast (the tropical Atlantic Ocean), brushing, or in some cases entering, the Carolinas or Florida, then "recurving" to sweep eastward back across the Atlantic. Let me emphasize one point right now, however. No place along the East or Gulf Coasts of the United States is or ever will be immune to a major hurricane. Despite what local folklore might say, there is no guarantee that a particular coastline will never be raked by the full fury of a great tropical storm. True, some areas have not been visited by a violent hurricane in many years, and coasts north of Cape Hatteras may be less threatened, statistically speaking, over the next few decades, but the threat is and never will be zero. It takes just one awesome storm a few hours to sweep away lives, fortunes, and dreams.

Hurricane statistics from the initial decades of the last 180-year cycle seem to support the theory. They show that a greater percentage of hurricanes then struck areas south of Hatteras than was the case during the surrounding decades. Table 3 shows that of all hurricanes recorded during the 30-year period 1800–1829 over half (55 percent) battered the Carolinas, Georgia, or the east coast of Florida. During 1770–1779 only 18 percent of the storms affected those coastlines, while in the 30 years from 1830 to 1859 the figure was just 28 percent. Table 3 also shows a reduced threat to areas north of Hatteras after 1800 (the start of the 180-year cycle) and suggests a slight reduction in the threat along the Gulf Coast between 1800 and 1829.

Remember, the percentages are for periods of 30 years. They don't tell us anything about the threat in any one particular year, which can be quite large or quite small. Also, because there was no organized system for observing and reporting hurricanes around 1800, and because population concentrations then were relatively sparse, more attention should be paid to the relative *change* in the percentage figures than to the actual percentages themselves.

These hurricane statistics from 2 centuries ago are based on the work of David Ludlum, who derived them from an exhaustive study of old records, diaries, and newspapers.[1] A point to keep in mind is that probably mention of only the strongest storms may have found its way into written records. Again, this stems from the fact that there was no formal reporting system and that reasonably dense population concentrations were few and far between.

An examination of hurricane statistics from 1880 through 1909, the beginning of the most recent long-term solar cycle, suggests pretty much the same thing as the earlier numbers. A U.S. Weather Bureau study completed in 1965 shows that a greater percentage of tropical storms and hurricanes recurved or remained in the westerlies (i.e., moved eastward) in the period 1886–1900 than during 1901–1963 (83 versus 63 percent).[7] A

TABLE 3 Percentage of total number of all recorded hurricanes affecting various U.S. Coastlines during the specified periods

	1770–1799	*1800–1829*	*1830–1859*
Eastport, Me. to Cape Hatteras, N. C.	53%	25%	23%
Cape Hatteras, N. C. to Key West, Fla.	18	55	28
Key West, Fla. to Brownsville, Tex.	29	20	49

Percentage of total number of all major or extreme hurricanes affecting various U.S. coastlines during the specified periods

	1850–1879	*1880–1909*	*1910–1939*
Eastport, Me. to Cape Hatteras, N.C.	27%	4%	8%
Cape Hatteras, N.C. to Key West, Fla.	27	46	34
Key West, Fla. to Brownsville, Tex.	46	50	58

tabulation of major or extreme* hurricanes reaching the U.S.
coast from 1880 through 1909 (Table 3) indicates that the percentage battering areas between Hatteras and Key West was
relatively higher than it was during the 30-year periods on either
side of 1880–1909. At the same time, the percentage ravaging
coasts north of Hatteras was relatively smaller. In the 3 decades
around the turn of the century no major or extreme hurricane
slammed into New England and only one hit north of Hatteras.

So far I've been talking about the *relative* frequencies of severe
hurricanes striking various sections of the U.S. coastline. I have
suggested that based on *percentages* the threat to areas north of
Hatteras over the next few decades will decrease, whereas south of
Hatteras it will increase. The actual number of hurricanes and
tropical storms probably will not be affected very much, at least
until later in the 1980–2009 period.

Hurricanes tend to form over warm tropical waters. Any
cooling of the tropics inhibits their formation. But Dr. Willett
theorizes that the cooling of the tropics and subtropics lags a
couple of decades behind that of the mid-latitudes.[4] Thus, the
total number of storms might be reduced slightly but not until
around the year 2000. Figure 13 shows the annual total of tropical
cyclones (hurricanes and tropical storms) from 1885 to 1976.
Notice that early in the period 1880–1909, the frequency of storms
didn't tail off for good until about 1895.

My review of the literature and statistics on hurricanes turned
up one interesting coincidence. At least I hope it is no more than a
coincidence, and no more than interesting. I assume it is a

*As defined in the definitive book *Atlantic Hurricanes*, by Gordon E. Dunn
and Banner I. Miller (Louisiana State University Press, 1964), an extreme
hurricane has winds exceeding 135 miles per hour and a minimum central
barometric pressure of 28.00 inches or less. A major hurricane has winds of
101–135 miles per hour with a central pressure of 28.01–29.00 inches. Only
major or extreme hurricanes are examined here because they may be more
analogous to the storms recorded around 1800 when the reporting system was
less than optimum (see previous comments in the text of this chapter).

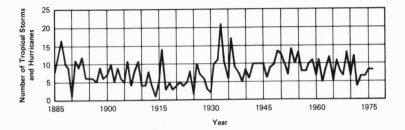

13. The annual variation in the number of tropical storms and hurricanes, 1885–1976. In the period 1880–1909, the frequency of storms did not diminish until about 1895. The suggestion is that the number of storms may again decrease but probably not until later in the 1980–2009 period, possibly around the year 2000.

statistical quirk that two of the most devastating hurricanes ever to slash through New England did so precisely 180 years apart, each 15 years after the start of a 100-year solar cycle: 1635 and 1815.

David Ludlum writes of the earlier storm: "Perhaps the historical stature of the 1635 event owes its prominence to the unusual severity of the storm itself, since the accounts of whole forests being leveled would indicate that it was a hurricane of exceedingly great force." Ludlum ranks it with only two other storms: the one of 1815 and the Great New England Hurricane of 1938.[1]

Of the 1815 storm Ludlum writes: "Of all the storms in New England's history the Great September Gale of 1815 was long accorded first place by local historians of the region." On Fisher's Island, off New London, Connecticut, wind and wave action swept an eight-square-mile piece of land clean of trees. A Mr. V. Utley of Lyme, Connecticut, kept a diary of the storm's destruction. In it he wrote: "The forest from New London to [the] Connecticut River . . . exhibits to the eye the most dreadful destruction ever made . . . whole forests of trees are either broken down, or torn up by the roots and crossing each other. . . ." Figure 14 charts the courses of the two great storms.

Another significant hurricane event took place in 1821, when a powerful tropical invader smashed across extreme western Long Island. The storm appears to have moved onshore near the present site of Kennedy International Airport. Available weather records indicate that over the past 200 years this was the only major hurricane to pass directly over a part of New York City.

So, although the *relative* hurricane threat to areas such as New England should diminish during the early decades of the new long-term solar cycle, it is obvious that some events of extreme violence have occurred in these same early decades in the past. I will not go so far as to predict that another superstorm will slash

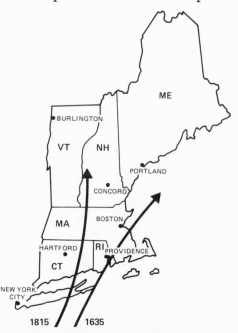

14. Paths of two of the greatest hurricanes ever to strike New England. They occurred precisely 180 years apart, each about 15 years after the start of a 100-year secular solar cycle. Whether this portends a recurrence of such an event around 1995 is problematical; on the other hand, it would perhaps be unwise to ignore the possibility entirely.

into New England or New York 15 years after the start of the next 100-year solar cycle, around 1995, but it seems a point worth keeping in the backs of our minds.

Probably a more important point worth remembering is one I made earlier. That was that no place along the Eastern Seaboard or Gulf Coast is or ever will be immune to a major hurricane. If you live near tidewater anywhere along the Atlantic or Gulf coasts, a few simple hurricane precautions seem essential. To the maximum extent practical make sure your home is "hurricane proof," have adequate insurance, and most importantly, evacuate your home—whether it is "hurricane proof" or not—when a warning to do so is given.

Evacuation to an area safe from the threat of tidal flooding (e.g., inland or higher ground) is the only way you can be *certain* of surviving a storm. Most hurricane-related deaths are due to drowning. The drownings occur when people are caught in huge storm tides and monstrous surf. The deaths are often even more tragic because many times they could have been avoided had people but heeded warnings and left their coastal homes before a storm tide inundated escape routes. Once escape routes are gone, people are at the mercy of water that in a severe hurricane may surge to twenty or more feet above mean sea level. Imagine trying to swim for your life in wrenching tidal currents, hundred-mile-an-hour winds, and floating debris.

Even after a hurricane has lost its powerful winds over land, it can still be a killer. "Camille," in 1969, after lashing the Gulf Coast with awesome two-hundred-mile-an-hour winds, quieted down after she moved inland over Mississippi. She took a peaceful journey through Tennessee and Kentucky, then unleashed torrential downpours on unsuspecting Virginia. Over a hundred people drowned as Camille poured out rains of up to thirty-one inches, most of it within a few hours. Valleys were flooded treetop tall in a deluge estimated to occur on the average of once in 1,000 years.

Three years later, in 1972, "Agnes," after losing intensity, dropped cloudbursts on New York State and Pennsylvania that produced unprecedented floods. Again, over one hundred people lost their lives as Agnes turned into the most costly storm, in terms of damage, in U.S. history. The bill for the storm totaled over three billion dollars.

The point to be emphasized here is that even after hurricane winds have calmed, the decaying circulation can hold immense amounts of water. Tremendous rains can occur even several days after a storm has moved inland. If you live in a deep valley or on a river plain susceptible to flooding, it is wise to stay apprised of the course of a dying hurricane even after it has been given last rites. Its death throes can be disastrous.

I suspect that drowning is an unpleasant way to die. In the case of hurricanes, or even ex-hurricanes, it is a way of death that in many cases can be avoided. The method is simple. Stay aware of the storm's track. Heed the warnings.

Tornadoes

There are perhaps no other natural phenomena in the United States that inspire as much fear and as much respect as tornadoes. The black, destructive twisters that commonly ravage the Midwest and South are most frequent in the spring. It is then when air mass contrasts are greatest: Warm, humid air streams northward from the Gulf of Mexico and collides with still chilly air from Canada. The result is an atmospheric battleground where death rides the sky in whirling funnels.

Tornadoes seem to occur at random and to be virtually unpredictable. Yet this in fact is not the case. The National Severe Storms Forecast Center (NSSFC) in Kansas City, Missouri, can, with surprising short-range accuracy, delineate areas where tornadoes are likely to form. These areas, or boxes, are designated as tornado "watch" regions. Once a tornado is actually detected, either visually or with radar, it is tracked continuously and warnings are issued.

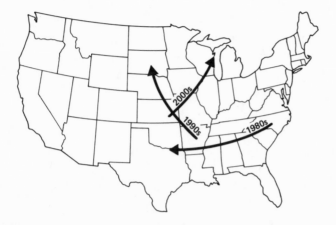

15. The arrows represent a schematic extrapolation of Fujita's discovery
that major, death-dealing tornadoes follow a cyclical pattern. The extrapolation suggests that the maximum threat of killer tornadoes during
the 1980s will move westward across the Deep South, shift northward
into the Great Plains in the 1990s, then head toward Wisconsin and
Illinois in the first decade of the twenty-first century. It's possible that
the transit of the westerlies to lower latitudes (Willett's theory) could
somewhat delay the shift of the maximum tornado threat to the more
northerly states. But the work of both Fujita and Willett would seem to
agree on one point: The 1980s will bring serious tornado problems to the
South.

Even on a long-range scale, tornado occurrence may not be
quite so random as we once thought. A recent study by T.
Theodore Fujita of the University of Chicago, Allen Pearson of the
NSSFC, and David Ludlum, concludes that major, severe tornadoes occur in a cyclical pattern.[8] At least this has been the case
back to 1911, which is as far back as the study goes. But the results
are very interesting. They indicate that the major twisters, the
ones which cause most of the deaths, are concentrated in only one
part of the United States at a time. Specifically, a pie-shaped area
of maximum risk appears to rotate slowly about a centroid located
in northeast Arkansas. One complete, clockwise rotation takes
about 45 years. Within the high-risk area, the death rate due to

tornadoes may be more than double what it is in other regions.

By extrapolating the rotation of the maximum risk quadrant, some ominious news for the Deep South appears. If the cycle continues, the maximum killer tornado threat in the 1980s will move from the Carolinas across Georgia, Alabama, and Mississippi into Louisiana and Arkansas. (See figure 15.)

Continuing the extrapolation suggests that during the 1990s the high-risk area will shift from Louisiana and Arkansas into northeast Texas and Oklahoma then northward through the Great Plains into Nebraska. During the first few years of the twenty-first century, areas in Wisconsin and Illinois may come under the gun.

The extrapolation of the cycle discovered by Fujita is consistent with what Willett's theory of the westerlies shifting somewhat southward would imply about tornadoes. Namely that the bulk of the tornado activity should occur in the Deep South during the 1980s and possibly even the 1990s. The westerlies should persist in their more southerly configuration even into the first decade of the new century. Thus, even then the bulk of major tornado activity might be confined to more southern locations, probably the high plains of Texas, Oklahoma, and Kansas. (The reason for this gets a little esoteric, so I will do no more than mention it deals with the positioning of undulations in the westerlies.)

Theory and extrapolation seem to agree on the 1980s, however: The Deep South may have to contend with a greater frequency of death-dealing twisters than will the rest of the country. As with hurricanes, the point should be made here that other areas of the country will not be free of the threat of severe tornadoes. All states east of the Rockies should continue to be prepared to deal with major tornado outbreaks, the South perhaps more so.

A brief word is perhaps in order regarding how the westerlies relate to tornadoes. In simplest terms, the westerlies mark the boundary, or zone of greatest contrast, between air masses. Given that the other meteorological ingredients necessary for tornado formation are present, the westerlies are conducive to tornado genesis because with their great store of temperature and moisture contrasts they possess a vast amount of potential energy. Even kinetic energy—energy derived from motion—may be present in the form of a jet stream wind. The more energy available, the more readily a tornado will form, and the more violent it will be.

There is some evidence that during 1880–1909, the period I am considering to be most analogous to what we are about to enter, the greatest percentage of large tornado outbreaks occurred in the southeastern states. Joseph P. Galway of the NSSFC carried out a study in which he showed that of the major tornado outbreaks (ten or more tornadoes occurring within certain geographical and time constraints) between 1880 and 1909, more slashed through the Southeast than through other areas of the country.[9] Admittedly, the figures may be somewhat biased because of different population distributions from those that exist now, but it's probably significant that the deadliest tornado swarms from around that turn-of-the-century period swept through areas south of the most recent headline makers.

For instance, in February 1884, a blitz of over sixty twisters killed an estimated 420 people from Alabama eastward to the Carolinas. Another massive tornado outbreak claimed over 270 lives and injured almost 1,400 people in April 1908. Louisiana, Mississippi, and Alabama took the brunt of that fury.[10]

Contemporary major tornado swarms have ravaged areas north of the Gulf Coast states. The Palm Sunday disaster in April 1965 snuffed out 257 lives in Indiana, Michigan, and Ohio. In April 1974 a huge outbreak of 148 twisters raked across the country from northern Alabama and northern Georgia into extreme southern Michigan. Over 300 people lost their lives to that uncontrolled display of nature's power. The most devastating tornadoes of that jumbo outburst leveled communities in Kentucky, Ohio, Indiana, and northern Alabama.

Because the South may become the primary stalking ground of killer tornadoes over the next couple of decades, does that mean if you live in Dixie, you should sell your home, pack your bags, and become a Yankee? In a word, no. As I mentioned earlier, tornado activity will still be widespread. Although major tornadoes may be more frequent in the Deep South, that doesn't rule out the possibility of North Dakota, Minnesota, or Massachusetts, for examples, having a severe outbreak.

The NSSFC does an excellent job of forecasting areas where tornadoes are likely to occur, and most communities in areas of high tornado frequency have excellent warning systems. But it is possible now in many locations to have an almost "personal" warning service. A large number of National Weather Service (NWS) offices broadcast weather forecasts and warnings continuously over special FM radio frequencies. The broadcasts are part of the NOAA (National Oceanic and Atmospheric Administration) Weather Radio System, which is designed to provide direct warnings into private homes for both natural disasters and nuclear attack. These warnings are intended to supplement those issued over commercial broadcast stations or sounded by local sirens.

The NOAA Weather Radio broadcasts usually use one of three high-band FM frequencies: 162.40, 162.475, or 162.55 megahertz (MHz). The broadcasts, most of which have a range of about forty miles, may be picked up by special low-cost receivers or by radios with special "weather bands." Some manufacturers offer weather radios that can alert you by sounding a tone or flashing a light that the NWS has issued a weather warning.

The government estimates that by 1980 about 90 percent of the U.S. population will be within listening range of a NOAA Weather Radio station.

Mobile homes, because they are relatively light, are notably susceptible to the vagaries of severe winds. If you live in a mobile home, you should pay particular attention to weather warnings of strong winds and tornadoes. One of the best ways you can be sure of receiving timely warnings is by investing in a weather radio, especially one that has the alerting capability. If you are concerned over the threat of tornadoes, you do not have to live in a bomb shelter or move to Point Barrow, Alaska. But you should know how to obtain up-to-the-minute weather warnings.

An interesting footnote to the study of tornadoes is that whereas tornadoes and their destruction grab all the headlines, it is lightning that kills more people annually in the United States than do either tornadoes or hurricanes.[11] You might remember that next time you are about to dash across the street during a thunderstorm.

Notes

1. Ludlum, D. *Early American Hurricanes*, (Boston: American Meteorological Society, 1963).
2. Fritts, H. "Tree-Ring Evidence for Climatic Changes in Western North America," *Monthly Weather Review* 93 (1965): 421.
3. Stockton, C. "Long-Term Spatial and Temporal Drought Frequency Analysis in Western United States Utilizing Tree Rings," NSF Report DES74-24163 (1976); and Stockton, C. and Meko, D. "A Long-Term History of Drought Occurrence in Western United States as Inferred from Tree Rings," *Weatherwise* 28 (1975): 245.
4. Willett, H. "Do Recent Climatic Fluctuations Portend an Imminent Ice Age?" *Geofisica Internacional* 14 (1974): 265.
5. _____. "Is U.S. Running Out of Water?" *U.S. News & World Report*, 18 July 1977, p. 33.
6. _____. *Climate, Climatic Change, and Water Supply* (National Academy of Sciences, 1977), p. 121.
7. Cry, G. "Tropical Cyclones of the North Atlantic Ocean" (Technical Paper No. 55), U.S. Government Printing Office, 1965.
8. Fujita, T.; Pearson, A.; and Ludlum, D. "Long-Term Fluctuation of Tornado Activities," SMRP Paper No. 128, The University of Chicago, 1975.
9. Galway, J. "Some Climatological Aspects of Tornado Outbreaks," *Monthly Weather Review* 105 (1977): 477.
10. Ludlum, D. *Weather Record Book*, (Weatherwise, Inc., 1971), pp. 10–13.
11. Mogil, H.; Rush, M.; and Kutka, M. "Lightning—A Preliminary Reassessment," *Weatherwise* 30 (1977): 192; and Mogil, H. and Groper, H. "NWS's Severe Local Storm Warning and Disaster Preparedness Programs," *Bulletin of the American Meteorological Society* 58 (1977): 318.

CHAPTER 4

The West Coast

STARTING WITH the West Coast and moving eastward, the next eight chapters discuss what the climate of the coming decades may be like in different regions of the United States. Specifically, the chapters deal with such meteorological phenomena as temperature, rainfall, and snowfall, and compare contemporary (1941–1970) means of those phenomena to averages that existed around the turn of the century. It is these turn-of-the-century averages rather than the current "normals" that may be more representative of what is in store for your area of the country in the immediate future.

Western Oregon and Western Washington

Perhaps my mother had the right idea. After enduring 30 years of Oregon rain she decided she and my father should move to a home in the Arizona sun. The Portland, Oregon, climate records (table 4) of the late 1800s and early 1900s would seem to support that decision with the implication that coming decades in the Pacific Northwest are going to be wetter and maybe a bit cloudier. As far as my father is concerned, that is good news. He is a steelhead fisherman. (A steelhead is a big sea-going trout that comes up the coastal rivers in the autumn when the rains start.)

Annual rainfall in the western parts of Oregon and Washington may increase on the order of 10 to 15 percent over what it's been the past 30 years. Table 4 suggests that most of that increase will come in the winter, the traditional monsoon season in the Northwest. The summers, perhaps the most beautiful in all of the United States, should escape pretty much unscathed: Little change from contemporary normals in either rainfall or temperature is suggested.

As a matter of fact, comparing 1880–1909 data with those of 1941–1970 suggests that temperatures will change very little even on an annual basis. At first I thought this might be the result of an

TABLE 4   Portland, Oregon, Monthly Averages

	Temperature		Precipitation[b]		Snow		Sunshine[c]	
	1880–1909 (1980–2009)?	1941–1970 [a]	1880–1909 (1980–2009)?	1941–1970 [a]	1880–1909 (1980–2009)?	1941–1970 [a]	1890–1903	1950–1973
Jan.	39.0°	38.1°	6.68″	5.88′	6.7″	4.9′	31%	23%
Feb.	41.5	42.8	5.15	4.06	3.9	0.8	36	36
Mar.	46.6	45.7	3.94	3.64	0.6	0.7	41	42
Apr.	51.7	50.6	3.10	2.22			43	49
May	57.3	56.7	2.29	2.09			49	54
June	61.9	62.0	1.60	1.59			69	50
July	66.9	67.1	0.57	0.47			61	70
Aug.	66.5	66.6	0.59	0.82			49	64
Sept.	61.3	62.2	1.74	1.60			44	58
Oct.	54.2	53.8	3.27	3.59			23	40
Nov.	46.1	45.3	6.12	5.61	0.3	0.3	20	28
Dec.	41.6	40.7	7.37	6.04	3.1	1.4	22	20
Annual	52.9°	52.6°	42.42″	37.61″	14.6′	8.1″	41%	47%

[a] Current normals.

[b] Includes melted snow.

[c] Percentage of total possible.

observing site location change. Contemporary observations are taken at the international airport, outside of the city center, where the older data were collected. The airport is near the mouth of the Columbia Gorge, which is a virtual sea-level cut through the Cascade Mountains to the east. The gorge is a potential source of colder air from interior regions in the winter and hotter air in the summer. You'll note in table 4 that current monthly average temperatures in Portland are usually a bit colder in the winter and slightly warmer in the summer than were turn-of-the-century readings.

But the same pattern emerged from my study of records from other cities located in the northwest quadrant of the United States. These other records are from sites that have not been shifted in a manner that would make them more vulnerable to hotter air in the summer and colder air in the winter. Particularly striking is the interstation similarity of some rather unusual changes in wintertime temperatures over the past century. At Portland, Oregon; Salt Lake City, Utah; Denver, Colorado; and Havre, Montana, Februarys from 1880 through 1909 show up as definitely colder when compared with contemporary normals; Decembers and Januarys come out milder (winters as a whole average out a bit milder—or not quite so cold). Such a well-defined, similar pattern would not occur over such a large area of the country by chance. Nor would it result from altered weather observing locations. Rather, it most likely would occur in response to a change in the preferred month-to-month configuration of the westerlies over a long time period (e.g., at least 30 years).

Thus, when we consider the wintertime temperature changes indicated by the Portland records as "real," and note the fact that average annual snowfall in the late 1800s and early 1900s was nearly double what it is now, the implication for coming winters in the Northwest becomes clear: certainly snowier, but with slightly milder Decembers and Januarys, and definitely colder Februarys.

The Pacific Northwest, at least west of the Cascades, is not really known as deep snow country, but that reputation could be changed over the next few decades. When we look at historical parallels in Portland, we see that in 1892-93 the seasonal snowfall added up to almost 61 inches and the snow depth reached a record 19 inches. The contemporary maximum snowfall is 44.5 inches set in 1949-50. Most stations in the Northwest did not keep records that would have encompassed the 1892-93 snowfall season, but several other records were set in the 1890s. Tatoosh Island, on the northwest tip of the state of Washington, set a twenty-four-hour snowfall mark with 14 inches in 1893. And Olympia, Washington, did the same with a 15-inch fall in 1895.

It is also interesting to note that in the period 1880-1909 there was no winter in Portland without measurable snow. Eight such winters occurred between 1941 and 1970. The suggestion, then, is that very soon sales of snow tires and sleds should pick up nicely in the lowlands of the Northwest.

A general increase in precipitation for Oregon and Washington, particularly in the winter, implies deeper snow in the mountains. That's good news for water supplies and winter sports. But there could be some real battles from time to time in keeping mountain passes open.

Heavy rainfalls could lead to occasional flooding problems, too. Again, considering parallels from the past, the gray skies of December 1882 let loose with over 20 inches of rain on Portland. That record has never been approached. But January 1953 did soak the city with 12.83 inches. Portland's record annual rainfall was established in 1882-83 (July through June): 72.89 inches. Seattle did not start keeping records until 1892, but they set their yearly rainfall record in 1893-94 with 45.73 inches.

The sunshine data for Portland during the 1890s and early 1900s look a little suspicious when considered on a month-by-month basis because of the large differences from current values. But averaged over a season (e.g., December-January-February for winter), the percentages appear reasonable, that is, the differences seem plausible. The most significant implication from the sunshine records is that autumns may be more cloud-shrouded in the near future.

Despite the suggestions that the Northwest is going to be wetter and snowier in coming decades, I will still consider it a great place to live. The summers have always been the redeeming season of the North Pacific states, and they should continue to be.

A bright, clear July day on Puget Sound with the wind snapping in from the Pacific and Mount Olympus pushing her snowy peak out of the rainforest to the west is worth a hundred rainy days. Well, at least ten.

Northern and Central California

The Pacific Northwest should not be alone in reaping the benefits of increased rainfall in coming decades. Records from San Francisco (table 5) suggest that precipitation over northern and central California will increase on the order of 5 to 10 percent in coming decades. Most of the increase in precipitation should occur in February and March with little change likely during the rest of the year. Summers will still be dry and sunny, superb for outdoor recreation.

Just how wet were the late 1800s and early 1900s in northern and central California? Witness the following records for maximum yearly precipitation: 64.47 inches at Eureka in 1904, 53.22 inches at Red Bluff in 1887-88, and 33.80 inches at Sacramento in 1889-90. Eureka also set a monthly precipitation mark in February 1902, when over a foot and a half (19.49 inches) of rain soaked the town. In January 1881 San Francisco must have seemed more like Seattle: A record 4.67 inches of rain cascaded down in one 24-hour period. Other 24-hour records were established at Fresno in November 1900 (2.86 inches) and Sacramento, April 1880 (7.24 inches).

As you might suspect, increased rainfall in lower elevations usually means increased snowfall in the higher elevations. Mount Shasta City fought off a record 270.5-inch seasonal snowfall in 1889-90; and a state record was established at Tamarack in the Sierra Nevada mountains in 1906-07, when almost 74 feet of snow fell. (It has never been clear to me how snowfalls of that magnitude are measured. I suppose that after the weather station is buried, it all becomes academic anyhow.)

In the same season (1889-90) that Mount Shasta City set its snowfall record, deep snows in the Sierras shut down operations on the Central Pacific Railroad for two weeks, almost double the length of any previous or subsequent stoppage. Railroad agents at Donner Summit measured 776 inches of snow that winter. Even

TABLE 5 San Francisco, California, Monthly Averages

	Temperature		Precipitation[b]		Snow[c]		Sunshine[d]	
	1880–1909 (1980–2009)?	1941–1970[a]	1880–1909 (1980–2009)?	1941–1970[a]	1880–1909 (1980–2009)?	1941–1970[a]	1895–1903	1936–1973
Jan.	49.8°	50.9°	4.21″	4.51″			53%	56%
Feb.	52.0	53.4	3.40	2.97			56	62
Mar.	53.4	54.3	3.42	2.77			60	69
Apr.	55.2	55.3	1.84	1.63			70	73
May	56.7	56.7	0.87	0.54			67	72
June	58.2	58.7	0.23	0.17			76	73
July	58.5	58.5	0.02	0.01			71	66
Aug.	58.9	59.4	0.01	0.05			60	65
Sept.	61.0	62.2	0.50	0.17			67	72
Oct.	59.9	61.4	1.27	1.06			67	70
Nov.	56.3	57.4	2.37	2.60			57	62
Dec.	51.2	52.0	4.05	4.18			56	53
Annual	55.9°	56.7°	22.19′	20.66′			63%	67%

[a] Current normals.

[b] Includes melted snow.

[c] Not significant.

[d] Percentage of total possible.

the Central Pacific's new rotary snowplow could not cope with the massive snowbanks; snowbound trains ended up being freed by an army of shovelers.[1]

Significant snow events around the turn of the century weren't confined to the mountains. Even in San Francisco you could have taken a little journey on cross-country skis in early February of 1887. A record snow of 3.7 inches covered downtown areas and up to 7 inches blanketed western hills. San Francisco's second greatest snowstorm in history came during the same decade. New Year's Eve afternoon of 1882 brought 3.5 inches of snow.

On the northern California coast at Eureka there have been only three noteworthy snows since 1886. Two of them occurred before 1909. In early March of 1896 a total of 2.9 inches coated the area. In January 1907 over 3 inches fell in 24 hours, and the month's total of 6.9 inches has never been equaled. Lest inland residents feel left out, Sacramento's greatest seasonal snowfall (4 inches) came in 1887-88.

Thus, records from 1880 through 1909 imply more snowfall for northern and central California over the next few decades. That bodes well for both water supplies and skiers. There will be some busy years for the big plows in the mountains, but that can be tolerated in view of the benefits.

San Francisco temperature records indicate that the decades around 1900 were a bit cooler than contemporary ones, although not strikingly so. The records suggest that during 1980-2009 the months from September through March might average out around a degree cooler, with the biggest differences from current means showing up in October and February. Very little temperature change is indicated for the summer months.

So the news for the northern two thirds of California is basically good. The decades in the immediate future should be slightly cooler and wetter, but not drastically so; adequate or better snows should blanket the mountains in most winters; and sunshine and warmth will continue to dominate the summers. (The ever-present coastal fog will still be around, of course.)

I wonder if you could cross the Golden Gate on cross-country skies.

Southern California

Records from Los Angeles suggest that southern California, too, will have somewhat greater rainfall over the next several decades. A 10-to-15 percent increase in annual precipitation is implied by the 1880–1909 figures (table 6). Most of the increase is likely from December through March.

Several precipitation maxima from that earlier era are worthy of note. At Los Angeles the maximum monthly rainfall record was

TABLE 6   Los Angeles, California, Monthly Averages

	Temperature		Precipitation[b]		Snow[c]		Sunshine[d]	
	1880–1909 (1980–2009)?	1941–1970[a]	1880–1909 (1980–2009)?	1941–1970[a]	1880–1909 (1980–2009)?	1941–1970[a]	1897–1903	1945–1976
Jan.	54.3°	56.7°	3.00″	3.00″			68%	69%
Feb.	55.4	58.1	3.11	2.77			73	72
Mar.	56.8	59.2	3.15	2.19			69	73
Apr.	59.5	61.7	0.99	1.27			70	70
May	62.3	64.7	0.49	0.13			60	66
June	66.5	68.0	0.08	0.03			67	65
July	70.4	73.2	0.02	0.00			78	82
Aug.	71.4	74.1	0.04	0.04			76	83
Sept.	69.4	72.7	0.10	0.17			76	79
Oct.	64.7	68.4	0.75	0.27			75	73
Nov.	60.5	62.7	1.34	2.02			79	74
Dec.	56.6	58.1	2.77	2.16			80	71
Annual	62.3°	64.8°	15.84″	14.05″			73%	73%

[a] Current normals.
[b] Includes melted snow.
[c] Not significant.
[d] Percentage of total possible.

set in December 1889, when 15.80 inches (over seven times the current monthly average) inundated the city. All-time annual maxima were measured at Los Angeles in 1884 (40.29 inches) and San Diego in 1883-84 (25.97 inches).

Another important precipitation event happened in the late 1800s: interior southern California's greatest snowstorm! On January 14, 1882, 5 inches of snow fell at Riverside, 15 inches piled up at San Bernadino, and up to 20 inches blanketed some interior mountains on the Mexican border. Snow covered the hills behind San Diego at El Cajon down to their bases.

Despite the fact that annual precipitation totals may increase over the coming decades, southern California may not be without significant drought problems, as suggested in the preceding chapter. During the years 1894 through 1904, for instance, annual precipitation totals at Los Angeles were all below the average of 15.84 inches.

Southern California may be slightly cooler as the twenty-first century approaches. The Los Angeles figures indicate that the period 1880 through 1909 was 2.5 degrees cooler, on an annual basis, than was 1941–1970. Some of that difference can be attributed to urbanization around the Los Angeles observing site during the later period (urban development tends to create a "heat island"—an area of warmer temperatures), but not all of it. Temperature records from Yuma, Arizona—several hundred miles to the east-southeast of Los Angeles on the California border—show a similar relatively large difference (1.7 degrees cooler during 1880–1909). Yuma's temperatures should be generally unaffected by urbanization. Records from both locations suggest that the month displaying the greatest cooling may be October (Octobers from 1880 through 1909 averaged over 3 degrees cooler than contemporary Octobers at both Los Angeles and Yuma).

The implications for southern California, then, are that the next several decades will not be quite so dry and so hot as recent ones. And, even though some very wet years are possible, some drought (especially around the beginning of the next century—see chapter 3) is also possible. There will still be some blistering hot months, of course, but Septembers and Octobers, in particular, should average out significantly cooler.

Sounds good. Now if we could just get rid of the damn smog.

Notes

1. Ludlum, D. "Early Railroading over the Snowy High Sierra," *Weatherwise* 26 (1973): 256.

CHAPTER 5

The Southwest

ALTHOUGH THE southwestern United States may become the prime target for the next major western drought (chapter 3), much of the area may get a bit more rainfall, on the average, over the next few decades than it did during the past several decades. Table 7 indicates that Yuma, Arizona, had almost 30 percent more annual rainfall during the period 1880–1909 than during 1941–1970. Records from Albuquerque, New Mexico (table 8), show very little change in precipitation totals between the two tridecade periods, but the older averages were computed from just 20 years' worth of data.

It may seem contradictory that the Southwest is likely to be threatened by a significant drought in the coming decades while at the same time weather records imply an increase in rainfall, particularly for Arizona. Yet it is not. Remember we are looking at 30-year averages. That means we are lumping drought years together with some relatively wet ones. And that it turn means in the averaging process the dry years are more than offset by some unusually rainy ones. (Perhaps the words *not so dry* rather than *rainy* should be applied to the arid Southwest.)

As an example of what I am talking about, consider the

following statistics from Yuma: The wettest year on record was 1905 (11.41 inches), the wettest month on record was August 1909 (6.25 inches), which also had the wettest 24 hours ever (4.01 inches); and the greatest number of consecutive days with measurable rainfall was 7, in 1897. Yet in the same 30-year period (1880–1909) there was a stretch of 351 straight days with no measurable rain (December 29, 1879, to December 15, 1880), a mark that still stands.

The year 1905 brought a record-breaking annual rainfall to other Arizona cities besides Yuma. Namely, Phoenix with 19.73 inches and Tucson with 24.17 inches.

"Damp" years, such as 1905, probably occurred in the complete

TABLE 7　Yuma, Arizona, Monthly Averages

	Temperature		Precipitation [b]		Snow [c]		Sunshine [d]	
	1880–1909 (1980–2009)[?]	1941–1970 [a]	1880–1909 (1980–2009)[?]	1941–1970 [a]	1880–1909 (1980–2009)[?]	1941–1970 [a]	1880–1909 [e]	1951–1976
Jan.	54.7°	55.4°	0.43″	0.38″				85%
Feb.	58.6	59.4	0.54	0.27				88
Mar.	63.5	63.9	0.40	0.21				91
Apr.	69.8	71.2	0.10	0.11				94
May	76.2	78.7	0.03	0.03				96
June	84.3	85.8	0.00	0.00				97
July	91.0	93.7	0.13	0.18				89
Aug.	90.3	92.8	0.52	0.44				92
Sept.	84.4	87.1	0.20	0.22				94
Oct.	72.8	75.9	0.23	0.27				92
Nov.	62.5	63.5	0.34	0.22				86
Dec.	55.9	56.3	0.53	0.34				83
Annual	72.0°	73.7°	3.45″	2.67′				91%

[a] Current normals.
[b] Includes melted snow.
[c] Not significant.
[d] Percentage of total possible.
[e] Figures not available.

absence of the semipermanent high-pressure area I talked about in chapter 3. In such a case the lower-latitude westerlies would be able to sweep moisture into Arizona off the Pacific, probably in the form of frequent frontal passages during the cool season, November through March. You will note (table 7) that it is in those months that most of the precipitation increase is implied.

Though the Albuquerque record indicates little change in yearly precipitation over the coming decades, it does imply a slight (16-percent) decrease in summertime (June through August) rain-

TABLE 8   Albuquerque, New Mexico, Monthly Averages

	Temperature		Precipitation [f]		Snow		Sunshine [h]	
	1893–1909 (1980–2009)? [p]	1941–1970 [a]	1889–1909 (1980–2009)? [p]	1941–1970 [i]	1891–1909 (1980–2009)? [p]	1941–1970 [a]	1880–1909 [a]	1940–1976
Jan.	34.8° [b]	35.2°	0.48″	0.30″	1.9″	1.9″		73%
Feb.	39.9 [c]	40.0	0.32	0.39	1.7	1.7		73
Mar.	47.6 [d]	45.8	0.28	0.47	0.6	1.8		74
Apr.	55.8	55.8	0.71	0.48		0.3		77
May	63.8	65.3	0.52	0.53				80
June	73.5	74.6	0.45	0.50				83
July	76.3	78.7	1.25	1.39				76
Aug.	74.5 [c]	76.6	1.02	1.34				76
Sept.	67.6	70.1	0.94	0.77				80
Oct.	56.3 [e]	58.2	0.66	0.79				79
Nov.	44.0 [e]	44.5	0.60	0.29	0.7	1.0		78
Dec.	34.9 [e]	36.2	0.45	0.52	1.5	2.8		72
Annual	55.8°	56.8°	7.68′	7.77′	6.4″	9.5″		77%

[a] Current normals.
[b] 1907 and 1909 missing.
[c] 1909 missing.
[d] 1904 missing.
[e] 1908 missing.
[f] Includes melted snow.
[g] 1891 missing.
[h] Percentage of total possible.
[i] Figures not available.

fall. Denver, Colorado, several hundred miles to the north, shows the same pattern.

Greater amounts of annual precipitation imply greater amounts of snowfall in the higher elevations of the Southwest, yet the Albuquerque statistics suggest just the opposite there: a 3-inch annual decrease. Yet it would not surprise me to see average annual snowfall totals creep upward over the next few decades, particularly in the mountains of Arizona.

Even the lower elevations might catch it once in a while. Las Vegas, Nevada, collected a foot of snow, a record, in December 1909. And Roswell, in southeastern New Mexico, measured a total fall of almost 31 inches in the winter of 1904-05. That amount has never been equaled. In that same winter Roswell residents shivered in the coldest weather ever there, when the mercury slumped to 29 below on February 13.

Yet the implications are that it will not be the winters in the coming decades that will experience the greatest cooling, but the hotter months. Winters may cool off a little bit, but it is May through October that will likely see the largest drop in mean temperatures. Averages in some months may decrease as much as 2½ degrees (July and September, Albuquerque) to 3 degrees (October, Yuma).

Not much, you say. Well, if Yuma cools off (is not so hot) from June through September to the extent suggested, the amount of electricity you use to run your air conditioner would be reduced on the order of 9 to 10 percent. That translates into a similar reduction on your electric bill, at least for that portion covering air conditioning.

As an antithesis to the cooling, a curious warming in March in the immediate future is implied by the Albuquerque figures. That could be a statistical quirk due to an incomplete set of temperatures. But the same magnitude of March warming is suggested by the stations I examined in a large area to the north of Albuquerque: Salt Lake City, Denver, and Dodge City, Kansas.

Thus, the suggested March warming is probably real, and probably covers a large area of the West from Utah through Colorado and New Mexico into western Kansas, at least.

But the bottom line is that through most of the year the Southwest should not be quite so hot in coming decades. And it should not be quite so dry—*on the average. On the average* is the disclaimer. It means the threat of extreme drought, probably along about the year 2000, cannot be dismissed.* Most years probably will be a bit cooler (1909 was the coolest year on record at Yuma) and a bit wetter than ones of the recent past. But there will still be years with long stretches of sweltering, parched weather.

That is the price that must be paid, I suppose, for sunshine and warmth. I only hope they do not become too costly.

*Chapter 3 discusses the drought problem in detail.

CHAPTER 6

The Interior West

THERE AREN'T many trees in Denver, Colorado. At least compared with western Oregon, where I was raised, and to New England, where I live. Still, Denver is one of my favorite cities. And I was hoping, as I studied old climate records from the area, that more precipitation would be implied for that region in the coming decades. A little more water, I reasoned, would encourage more trees to take up residence in eastern Colorado. Unfortunately, it is the "Big Sky" effect that will prevail. "Big Sky" is a High Plains euphemism for "no trees."

The records from Denver (table 9) suggest that that area will be one of the few in the United States to experience a decrease in precipitation over the next 20 to 30 years. Not only that, of all the records I studied, Denver's suggested the largest drop in average annual precipitation: over 9 percent, or 1½ inches, which is significant for a place that receives only 15½ inches each year.

The suggested decrease in precipitation should be limited to the east slope of the central Rockies, however. Records from Salt Lake City, Utah, and Havre, Montana, imply increased yearly rainfall totals for the remainder of the Interior West. (I'm defining the Interior West as extending from the Cascade and Sierra Nevada

TABLE 9   Denver, Colorado, Monthly Averages

	Temperature		Precipitation [b]		Snow		Sunshine [c]	
	1880–1909 (1980–2009)?	1941–1970 [a]	1880–1909 (1980–2009)?	1941–1970 [a]	1882–1909 (1980–2009)?	1941–1970 [a]	1890–1903	1950–1976
Jan.	30.7°	29.9°	0.45″	0.61″	5.1″	8.6′	73%	72%
Feb.	31.7	32.8	0.52	0.67	6.4	8.1	67	71
Mar.	38.4	37.0	1.02	1.21	9.6	13.5	67	70
Apr.	47.8	47.5	2.16	1.93	9.5	9.2	68	66
May	56.3	57.0	2.36	2.64	1.8	1.9	61	64
June	66.1	66.0	1.46	1.93			69	70
July	71.6	73.0	1.64	1.78			67	71
Aug.	70.9	71.6	1.31	1.29			69	72
Sept.	62.9	62.8	0.94	1.13	0.7	1.2	75	74
Oct.	51.0	52.0	1.03	1.13	4.0	4.0	76	73
Nov.	39.4	39.4	0.55	0.76	4.7	7.9	71	66
Dec.	33.3	32.6	0.58	0.43	7.1	5.6	68	69
Annual	50.0°	50.1°	14.02″	15.51″	48.9′	60.0′	69%	70%

[a] Current normals.
[b] Includes melted snow.
[c] Percentage of total possible.

mountains eastward through Montana, Wyoming, and Colorado.)

On an annual basis, little change in temperature is indicated for most of the Interior West over the next several decades. However, cooler monthly averages are likely over eastern Montana. Records from Havre (table 10) show that during 1880–1909, the period May through November was significantly cooler than contemporary means indicate.

In general, winters should be marginally milder across much of the West from 1980 through 2009, and summers a bit cooler. At Denver (table 9) and Salt Lake City (table 11) the largest drop in monthly average temperature is indicated for July. Winter temperature averages from around the turn of the century at Havre,

Denver, and Salt Lake City show the same pattern of departures from current means as do the figures from Portland, Oregon. That is, around 1900 Decembers (especially) and Januarys were often milder (or not so cold), and Februarys colder, than they were from 1941 to 1970. Thus, the implication for the next few decades is that whereas Decembers and Januarys will not be quite so cold as recently—on the average—Februarys will frequently be more wintery. This pattern may be particularly prominent around Havre, where Februarys from 1880 to 1909 averaged almost 4

TABLE 10   Havre, Montana, Monthly Averages

	Temperature		Precipitation[d]		Snow		Sunshine[e]	
	1880–1909 (1980–2009)?	1941–1970[a]	1880–1909 (1980–2009)?	1941–1970[a]	1884–1909 (1980–2009)?	1941–1970[a]	1880–1909[f]	1961–1976
Jan.	12.3°[b]	11.3°	0.73′	0.52″	6.8″	7.8″		49%
Feb.	13.9 [b]	17.6	0.50	0.40	5.1	5.9		60
Mar.	26.8 [b]	26.5	0.54	0.49	5.7	5.6		69
Apr.	43.8 [b]	43.6	0.89	1.02	2.9	5.4		67
May	53.2 [b]	54.9	2.04	1.48	2.5	0.2		71
June	61.3 [c]	62.1	2.93	2.55		0.1		73
July	68.0 [b]	69.9	1.89	1.38				85
Aug.	66.1	68.0	1.21	1.05				82
Sept.	55.9	57.1	1.06	1.11	0.1	0.4		74
Oct.	44.8	46.6	0.62	0.67	2.0	1.5		63
Nov.	29.3	30.0	0.68	0.46	5.1	5.5		49
Dec.	21.1	18.2	0.54	0.42	5.0	5.9		44
Annual	41.4°	42.2°	13.63″	11.55″	35.2′	38.3′		68%

Note: Until 1892 observations were taken at Fort Assinniboine, Montana, seven miles southwest of Havre.

[a] Current normals.
[b] 1880 missing.
[c] 1880, 1881, and 1885 missing.
[d] Includes melted snow.
[e] Percentage of total possible.
[f] Figures not available.

TABLE 11 Salt Lake City, Utah, Monthly Averages

	Temperature		Precipitation [b]		Snow		Sunshine [d]	
	1880–1909 (1980–2009)?	1941–1970 [a]	1880–1909 (1980–2009)?	1941–1970 [a]	1884–1909 (1980–2009)?	1941–1970 [a]	1890–1903	1938–1976
Jan.	28.9°	28.0°	1.29′	1.27″	10.8″ [c]	13.4″	43%	47%
Feb.	32.9	33.4	1.50	1.19	10.3 [c]	9.7	44	55
Mar.	41.1	39.6	1.97	1.63	10.2 [c]	10.8	52	64
Apr.	49.8	49.2	2.07	2.12	2.5 [c]	5.2	59	66
May	57.4	58.3	2.06	1.49	0.5 [c]	0.6	64	73
June	66.9	66.2	0.79	1.30			79	78
July	75.3	76.7	0.42	0.70			81	84
Aug.	74.4	74.5	0.89	0.93			77	83
Sept.	64.3	64.8	0.89	0.68		0.1	79	84
Oct.	52.1	52.4	1.32	1.16	0.8 [c]	0.8	69	73
Nov.	40.6	39.1	1.31	1.31	6.0	6.2	55	55
Dec.	32.2	30.3	1.38	1.39	9.5	12.4	44	45
Annual	51.3°	51.0°	15.89′	15.17″	50.6″	59.2″	62%	70%

[a] Current normals.
[b] Includes melted snow.
[c] 1885–1909.
[d] Percentage of total possible.

degrees colder than contemporary ones.

Milder readings for March, as well as for December, are implied by the 1880–1909 data. As you will see in succeeding chapters, the significant March warming is a phenomenon that may soon appear not only in the intermountain West and across the Rockies but through the southern Great Plains and eastward all the way to the Appalachians as well.

Eastern Colorado and Eastern Wyoming

The sinking of the westerlies to lower latitudes over the next few decades will cause increasing precipitation over most of the United

States. Unfortunately, the effect will be just the opposite for those
areas immediately in the lee (on the east side) of the Rocky Mountains. There a *rain shadow* will prevail. A rain shadow is
produced when moisture-bearing winds (in this case the
westerlies) deposit their rain and snow on the windward (western)
side of a mountain range and then warm up and dry out as they
descend in the lee of the mountains. The result is greatly
diminished precipitation in the lee of the mountains—a rain
shadow.

In the case of the Colorado and Wyoming plains adjacent to the
Rockies, the rain shadow will be enhanced by another factor. The
westerlies, as they move southward, will more frequently cut off
surges of moisture trying to move into that area from the Gulf of
Mexico. Denver records (table 9) suggest that 9 of the 12 months
each year will average out drier over the next 3 decades. An in-
significant precipitation increase is suggested for August, with
more important increases possible in April and December.

Rainfall in May and June may decrease on the order of 16 percent. That is bad news for wheat farmers in eastern Colorado
because the precipitation in May and June is extremely important
in determining how well their crops will do. But of even more importance is the May temperature.[1] The Denver figures imply that
May will average slightly cooler in the near future, and that
should have a positive effect on wheat crops.

Although precipitation totals may be reduced in eastern
Colorado and eastern Wyoming as the twenty-first century approaches, major drought should not be a threat. The dendrohydrology studies that I talked about in chapter 3 suggest that
the next significant western drought will probably be centered in
Arizona. Still, some very dry months will likely show up in the
eastern sections of Colorado and Wyoming. The driest recorded
month Denver ever had was December 1881, when not a drop of
rain nor a flake of snow fell.

The dry months may be offset by some extremely wet ones,

however. Both Cheyenne, Wyoming, and Pueblo, Colorado, experienced their wettest month on record in April 1900. Cheyenne measured 7.66 inches of precipitation, and Pueblo, 8.31 inches.

Temperatures, except in February, will probably average a little milder through the snowfall season in coming decades. That, together with less precipitation, suggests less snowfall. The implied decrease at Denver is over 10 inches a year!

Despite the suggested decrease in snowfall *averages*, some occasional record-breaking snowfalls may plague the area, thanks primarily to April blizzards. Denver was buried under almost 2 feet of snow in 24 hours in April 1885. Cheyenne established a single-storm record in April 1901, when 19.2 inches of snow piled up. And Cheyenne's snowiest month ever was April 1905, when 46.5 inches of snow turned southeastern Wyoming into a springtime winter wonderland. Both Denver and Cheyenne had record seasonal snowfalls in the early 1900s. Denver dug out from under 118.7 inches in 1908-09, and Cheyenne battled 100.8 inches in 1904-05.

Although snowfalls, on the average, may soon decrease on the plains adjacent to the Rockies, snowfalls in the mountains probably will not. Plenty of moisture should be carried eastward from the Pacific Ocean into the mountain West during the cold season. Salt Lake City records (table 11) suggest about a 10-percent increase in average precipitation from October through March for the coming decades. Mountain snowfalls would probably increase by a similar amount. A place in the Colorado Rockies, Ruby, set a state record for snowfall in the season of 1896-97 with a massive total of 65 feet. Ruby also established a single-month state record in March 1899: 249 inches. So, if you are a winter devotee of Aspen or Snowmass or Vail, don't despair.

The fact that most winters may not be quite so cold in coming decades should not preclude some very cold ones from occasionally occurring. Denver's coldest winter ever was in 1898-99 (24.6 degrees), and Colorado Springs established its all-time record low temperature in January 1883: a frigid 32 below.

Summers, while they may often be a little less hot in the near future, will still have the potential to produce some blistering heat. The hottest temperature on record for Colorado was measured on July 11, 1888, at Bennett, about 25 miles east of Denver. It hit 118 degrees.

It becomes apparent, then, that while drier weather with more moderate summers and winters may be the rule for the eastern sections of Colorado and Wyoming over the next few decades, extreme contrasts of an opposite nature will be possible on any individual day or during any one season. The TV weathermen will not run out of things to talk about.

Eastern Washington, Eastern Oregon, Idaho, Montana, and Northwestern Wyoming

Over the next few decades the area from the Cascade Mountains eastward through Idaho and Montana should experience a generous increase in precipitation. Records from Havre, Montana (table 10), suggest an 18-percent increase in average annual precipitation there. Figures from Portland, Oregon, imply about a 13 percent increase; apparently this increase will extend inland across the Pacific Northwest all the way through Montana.

Precipitation during the months of May, June, and July is quite important to crops in eastern Montana.[2] The Havre records indicate that a significant increase in precipitation during those months will take place in coming decades. That should benefit Montana farmers.

That the decades around the turn of the century were not nearly so dry as recent decades is supported by the large number of maximum precipitation records that were established. Havre's maximum yearly precipitation occurred in 1884: 25.67 inches. The maximum monthly precipitation record was set in July 1884: 9.67 inches. And the maximum 24-hour rainfall record was established in a June 1887 deluge: 3.71 inches.

Other all-time marks for maximum annual precipitation were set in Walla Walla, Washington (1893-94); Pendleton, Oregon (1899-1900); Pocatello, Idaho (1909); and Great Falls, Montana (1909). Monthly marks were numerous, too: Pendleton (November 1897); Boise, Idaho (March 1904); Pocatello (March 1907); and Kalispell, Montana (November 1897).

Not all of the years between 1880 and 1909 had plentiful rain and snow, however. Both Havre and Great Falls, Montana, had their driest year ever in the early 1900s: Great Falls in 1904 with only 6.68 inches of precipitation, and Havre in 1905 with but 6.76 inches. So, although most years in the near future should have adequate precipitation, a few dusty stretches are probable, too.

More precipitation implies more snowfall. But the Havre records suggest otherwise. However, because virtually all of the wintertime precipitation falls as snow there, I find it hard to believe that more precipitation could result in less snow. Using contemporary ratios of average monthly precipitation to average monthly snowfall as a basis for calculation, I would be willing to bet that Havre, in coming decades, will see an increase in average snowfall on the order of 10 inches per year.

Havre set a number of snowfall records in the late 1800s. The greatest seasonal snowfall total was 82 inches in 1889-90. That was set after an astounding May snowstorm polished off the season by dumping almost 25 inches on the city in 24 hours (a record). The all-time mark for maximum depth (32.2 inches) was set in that same snowy May. Havre's greatest single snowstorm blanketed the area with 28 inches in January 1884.

Other areas were snowy, too. Residents of Helena, Montana, watched a record 112.8 inches of snow sift down in 1880-81. In that same winter Helena established its all-time monthly mark with 46.4 inches in December. Boise, Idaho, set a monthly record in December 1884 with just over 3 feet of snow. Boise also measured 22 inches on the ground that month, a depth that has never been equaled there.

While eastern Montana may be somewhat cooler, especially from May through October, in coming decades, the remainder of the northern Interior West will probably experience little overall temperature change. I would suspect, though, that summers might average out a little less hot, and winters a bit milder. The implication for individual months is that Marches and Aprils should be milder, and Februarys colder, than contemporary means indicate.

Extremes will still be possible, of course. Spokane, Washington, endured its all-time record low temperature in January 1888, when the mercury bottomed out at 30 below. Less than 2 decades later, in 1906, Spokane set a record for monthly heat with an average July temperature of 75.9 degrees. State records for high temperatures were set at Pendleton, Oregon, in August 1898 (119 degrees) and at Basin, Wyoming (in northwestern Wyoming), in July 1900 (114 degrees).

Overall, though, the northern Interior West should experience more moderate temperatures during the next 2 or 3 decades. Do not let that outlook dull your snow-shoveling instincts, though.

Nevada, Utah, Southwestern Wyoming, and Western Colorado

Data from Salt Lake City (table 11) indicate that much of the intermountain West and Rockies will be a bit wetter in the near future. Summers should average out drier, and springs wetter and cloudier, than has been the case in recent decades.

The Salt Lake City precipitation records generally support the conclusions of a comprehensive research project that dealt with secular changes in Rocky Mountain precipitation. The project was carried out by Raymond S. Bradley of the University of Massachusetts. Bradley studied numerous precipitation records from Idaho, Montana, Wyoming, Utah, and western Colorado.

He reached the conclusion that summers, and in many spots autumns, in those areas were much drier 80 to 100 years ago than they were during 1941–1970. Winters and springs, Bradley decided, were wetter around the turn of the century.[3]

Extreme precipitation marks from Nevada eastward indicate the same variability during 1880–1909 that records from other parts of the West show. For instance, Reno, Nevada, had its wettest year (13.73 inches) in 1890, and Grand Junction, Colorado, had its driest year (3.64 inches) in 1900. Records for the wettest month ever were set at Lander, Wyoming, in April 1900; Grand Junction in September 1896; and Winnemucca, Nevada, in March 1884.

The implied general increase in precipitation, especially during the winter, means that deeper mountain snows are likely in the near future. That should make everybody happy. More snow suggests increased water supplies for much of the West, good skiing, and lots of winter tourists.

The Salt Lake City figures indicate a slight decrease in annual snowfall totals over the coming decades. This may be due to two factors: the slight warming implied for much of the winter, and the fact that the 1880–1909 measurements were made in the city, farther away from the Great Salt Lake than the contemporary observation site at the airport. The Great Salt Lake, which doesn't freeze in the winter, serves as a limited moisture supply in certain snowfall situations: Resultant snows decrease with distance from the lake shore. Mountain regions should be unaffected by the factors influencing snowfall in Salt Lake City, though. And copious snows should be the rule over the next 20 or 30 years in the high country. Skiers will still be able to go schussing through the deep powder at Alta and Snowbird.

Turn-of-the-century temperature averages suggest that in coming decades areas around Salt Lake City will be a bit milder from November through April, Februarys excepted. Summers should not average quite so hot as in the recent past, with Julys showing the most noticeable cooling.

Heat waves will probably still smother the region from time to time, though. Saint George, Utah, reached 116 degrees, a state record, in June 1892. And Elko, Nevada, baked in 107-degree heat one July day in 1890—the hottest day ever there. On the other end of the scale, Reno, Nevada's thermometer reached its nadir in January 1890: 19 below.

The bottom line for much of the intermountain West and Rockies for the period 1980–2009 is that most winters will not be quite so cold as in recent decades, but they will be damper and snowier; springs will average out milder, cloudier, and wetter; but the majority of summers should be super with less shower activity, plenty of sunshine, and a shade cooler temperatures, particularly in July; autumns may average slightly wetter and milder in Nevada and Utah but should be drier in western Colorado and western Wyoming (Bradley's study).

Climatically speaking, I would suspect that Utah, and surrounding areas, would be pretty good places to live during the next 20 or 30 years. Not that they are not now, of course.

Notes

1. Personal communication with Dr. James McQuigg, formerly director of the Center for Climatic and Environmental Assessment, February 1978.

2. _____. Local Climatological Data, Havre, Montana, 1976, National Climatic Center, 1977.

3. Bradley, R. "Secular Changes of Precipitation in the Rocky Mountain States," *Monthly Weather Review* 104 (1976): 513.

CHAPTER 7

The Great Plains

OVER THE NEXT 2 or 3 decades the Great Plains will probably be a little cooler and a little wetter. Generally speaking, that is good news for farmers. Of course, there will continue to be some excessively wet years and some very dry ones, but in many years both moisture and temperature should be conducive to excellent crop yields. Perhaps the most significant negative factor as far as agriculture is concerned will be the possibility of early killing frosts.

Most people, though, will not be concerned with early frosts. Rather, they will be more concerned with the costly fact that cooler weather will mean higher heating bills.

The Northern Plains

Although Minnesota is one of my favorite states, the idea of having to contend with Arctic temperatures in the winter has always stayed my enthusiasm for wanting to move there. Records from the northern Great Plains suggest that winters in the coming years will do even more toward keeping me a mere visitor to the Land of

83

84   WEATHER WATCH

Ten Thousand Lakes. And I will do my visiting in the summer, thank you.

Records from Bismarck, North Dakota (table 12), and Minneapolis–Saint Paul (table 13)—yes, I know they are separate cities, but I have drawn on data from both locations—indicate that winters were noticeably colder around the turn of the century. This is especially apparent in Bismarck's figures. The most frigid winter on record there was that of 1886-87, when the mean temperature for December, January, and February hovered just under zero (–0.6 degrees). North Platte, Nebraska's coldest winter ever (16.3 degrees) came in 1884-85.

In the Twin Cities of Minnesota there have been 9 months since

TABLE 12   Bismarck, North Dakota, Monthly Averages

	Temperature		Precipitation b		Snow		Sunshine e	
	1880–1909 (1980–2009)?	1941–1970 a	1880–1909 (1980–2009)?	1941–1970 a	1884–1909 (1980–2009)?	1941–1970 a	1897–1903	1940–1976
Jan.	7.0°	8.2°	0.49″	0.51″	5.6 c	7.3″	51%	54%
Feb.	9.1	13.5	0.48	0.44	4.9 d	6.3	56	56
Mar.	22.9	25.1	0.94	0.73	8.1 d	7.9	53	60
Apr.	42.7	43.0	1.50	1.44	2.6 d	4.0	56	59
May	54.2	54.4	2.23	2.17	1.2 d	1.3	59	63
June	63.8	63.8	3.48	3.58			60	64
July	69.3	70.8	2.19	2.20			69	76
Aug.	67.8	69.2	1.86	1.96			66	74
Sept.	58.0	57.5	1.07	1.32	0.1 d	0.3	63	66
Oct.	45.0	46.8	0.99	0.80	0.7	1.1	60	59
Nov.	27.5	28.9	0.67	0.56	6.0	4.8	49	44
Dec.	15.3	15.6	0.58	0.45	5.9	6.0	46	47
Annual	40.2°	41.4°	16.48″	16.16′	35.1″	39.0″	57%	62%

a Current normals.
b Includes melted snow.
c 1886–1909.
d 1885–1909.
e Percentage of total possible.

1819 that have averaged below zero. Four of those months occurred in the 1880s (January 1883, 1885, 1887, and 1888).[1]

Februarys on the northern Plains from 1880 through 1909 averaged markedly colder than contemporary Februarys. They were so bitter that state records for extreme cold were established in Nebraska in February 1899 (47 below) and in Minnesota in February 1903 (59 below). Perhaps you can now understand why I want to be only a summertime tourist in Minnesota.

Many all-time record low temperatures were set in February 1899, when a tremendous mass of Arctic air flooded the entire country east of the Rockies. The mercury was driven to sub-zero readings as far south as northern Florida, where Tallahassee

TABLE 13   Minneapolis–Saint Paul, Minnesota, Monthly Averages

	Temperature		Precipitation b		Snow		Sunshine e	
	1880–1909 (1980–2009)?	1941–1970 a	1880–1909 (1980–2009)?	1941–1970 c	1884–1909 (1980–2009)?	1941–1970 c	1896–1903	1939–1972
Jan.	12.0°	12.9°	0.93″	0.73″	7.8″ d	8.2″	49%	50%
Feb.	14.8	17.2	0.90	0.84	6.8 d	7.9	55	57
Mar.	28.3	29.0	1.48	1.68	8.6 d	10.4	49	55
Apr.	46.0	45.8	2.48	2.04	3.9 d	2.4	58	55
May	57.4	57.8	3.42	3.37	0.3 d	0.2	55	58
June	67.3	67.8	4.14	3.94			59	62
July	71.8	72.9	3.55	3.69			66	70
Aug.	69.6	71.1	3.17	3.05			59	67
Sept.	61.0	60.6	3.40	2.73		0.1	61	61
Oct.	48.6	50.7	2.43	1.78	0.2	0.4	52	57
Nov.	31.8	33.0	1.29	1.20	5.0	5.2	44	38
Dec.	19.3	19.4	1.08	0.89	6.4	8.6	44	40
Annual	44.0°	44.9°	28.27″	25.94″	39.0′	43.4″	54%	58%

a Actual, unadjusted means.

b Includes melted snow.

c Current normals.

d 1885–1909.

e Percentage of total possible.

registered 2 below. In South Dakota Rapid City (34 below) and Sioux Falls (42 below) tallied their lowest readings ever. In Nebraska North Platte (35 below) and Scottsbluff (45 below) did the same.

Other minimum marks around the turn of the century were set at Pierre, South Dakota (40 below, February 1905); Lincoln, Nebraska (29 below, January 1892); Omaha, Nebraska (32 below, January 1884); Des Moines, Iowa (30 below, January 1884); and in Minnesota at International Falls (49 below, January 1896), Rochester (42 below, January 1887), and Saint Cloud (42 below, January 1904).

Not only were winters colder, but the months from July through November, September aside, were significantly cooler 80 to 100 years ago. At both Bismark and Minneapolis–Saint Paul, October averaged about 2 degrees colder then. And even though most Septembers were slightly milder during that same time, record early freezes occurred on September 18, 1901, at both Pierre, South Dakota, and Omaha, Nebraska.

Bismarck's earliest freeze occurred in July! July 6, 1884. And the latest freeze in history there was on June 7, 1901. If both those events had taken place in the same year, the growing season would have been mighty short.

Thus, the implication for the next several decades is that temperatures on the northern Plains are going to be cooler, with colder winters, and particularly colder Februarys. Unusually early frosts may plague farmers from time to time, though the cooler temperatures should generally enhance crop yields.

Precipitation may increase slightly in coming years. The Bismarck data suggest about a 2-percent annual increase, and the Minneapolis–Saint Paul figures hint at roughly a 9-percent increase. A number of stations set yearly precipitation records around the turn of the century, indicating that that time was indeed damp. Williston, North Dakota, checked in with 23.35 inches in 1880, Saint Cloud, Minnesota tallied 41.01 inches in

1897, and in Iowa, Des Moines measured 56.81 inches in 1881, while Sioux City caught 41.10 inches in 1903.

A number of spots on the northern Plains set records for monthly wetness during 1880–1909. Tecumseh, Nebraska, established a state mark in June 1883, when 20 inches soaked the town. And Alexandria, Minnesota, set a record for that state in August 1900 with 15.52 inches of rain. Other monthly highs during the same period were set in Fargo, North Dakota; Rapid City and Sioux Falls, South Dakota; Minneapolis, Rochester, and Saint Cloud, Minnesota; and Des Moines and Sioux City, Iowa.

There were some serious drought years, too. Both Pierre, South Dakota (7.82 inches), and Valentine, Nebraska (10.14 inches), had their driest year ever in 1894. But indications for the coming decades are that it is the southwestern United States that may be the hardest hit by severe drought (see chapter 3).

The records from Bismarck and Minneapolis–Saint Paul suggest that annual snowfall will diminish a little over the next 20 or 30 years. In reality that would seem unlikely if temperatures decrease and precipitation increases. I would think it more likely that annual snowfall totals at both Bismarck and the Twin Cities might be 3 or 4 inches greater. Reports of some rather hefty snows contained in the older records support that idea. State marks for the snowiest season ever were established in North Dakota at Pembina in 1906-07, when just under 100 inches fell, and in Iowa at Northwood in 1908-09, when 90.4 inches came down.

In North Dakota, Fargo and Williston had their snowiest months ever in 1896. Fargo collected 30.4 inches in November, and Williston, 26.6 inches in April. Des Moines, Iowa, had a total snowfall of over 3 feet in January 1886, the snowiest month on record there.

Maximum depth marks were established at Fargo (27.8 inches) and Williston (27.1 inches) in March of 1897. And Grand Island, Nebraska, measured 28.4 inches on the ground, the deepest ever, in February 1905.

Limited data from Bismarck and Minneapolis–Saint Paul suggest that the coming decades might be slightly less sunny. However, any decrease in the percentage of possible sunshine might be noticeable only in March, and again in July and August.

Summing it up then, the climate information from 1880 through 1909 suggests that the northern Plains are going to be cooler and slightly wetter in the near future. Winters should be colder, especially in February; and the period July through November—September excepted—should also see a drop in monthly mean temperatures (Septembers may be slightly milder). A small increase in average annual rainfall and snowfall should occur, accompanied by a slight increase in cloudiness.

If the fact that winters may average a degree or two colder doesn't seem worth worrying about, let me put it another way: Residents of Bismarck may see their heating bills for the period October through March increase by an average of 4 or 5 percent; Twin City dwellers might expect to shell out a little over 3 percent more for the same period. Increases in any one individual year could be as great as 15 percent! Remember, those estimated increases in cost are based only on climatic factors and do not allow for increases from any other sources. Of course, increases from other sources will most certainly come, and they will most certainly be substantial.

Just add the cost fostered by climatic change onto them.

The Southern Plains

The story for the southern Great Plains is pretty much the same as that for the northern Great Plains. The next few decades should be cooler and slightly wetter. Figures from Dodge City, Kansas (table 14), imply about a 1-degree decrease in average annual temperature and about a 3-percent increase in average annual precipitation.

Winters should be colder, with Februarys showing the biggest departures from contemporary normals. The coldness of Februarys 80 to 100 years ago is exemplified by Dodge City's setting of its all-time record minimum, 26 below, in that great February 1899 Arctic outbreak. In that same super cold wave records were established in Topeka, Kansas (25 below); Wichita, Kansas (22 below); Oklahoma City, Oklahoma (17 below); and in Texas at Amarillo (16 below) and Fort Worth (8 below). Concordia, Kansas, had its bitterest day in January 1884, when the thermometer registered 25 below. The Kansas state record for cold was set in February 1905, when the temperature in Lebanon plunged to 40 below.

A number of spots on the southern Plains had their most frigid winter ever during the 1880–1909 period. Dodge City's coldest (22.0 degrees) came in 1884-85. Oklahoma City's chilliest (31.7

TABLE 14   Dodge City, Kansas, Monthly Averages

	Temperature		Precipitation [b]		Snow		Sunshine [c]	
	1880–1909 (1980–2009)?	1941–1970 [a]	1880–1909 (1980–2009)?	1941–1970 [a]	1885–1909 (1980–2009)?	1941–1970 [a]	1890–1903	1943–1976
Jan.	29.2°	30.8°	0.47″	0.50″	3.9″	3.7″	65%	68%
Feb.	31.7	35.2	0.73	0.63	5.4	3.4	63	64
Mar.	42.4	41.2	0.81	1.13	3.7	5.4	65	66
Apr.	54.2	54.0	1.89	1.71	0.9	0.6	66	69
May	63.1	64.0	3.18	3.13			64	70
June	72.9	73.7	3.65	3.34			71	76
July	77.4	79.2	3.32	3.08			74	79
Aug.	76.6	78.1	2.46	2.64			79	77
Sept.	68.8	68.9	1.81	1.67			75	74
Oct.	56.2	57.9	1.58	1.65	0.2	0.2	76	74
Nov.	42.0	42.8	0.80	0.59	1.3	2.3	66	67
Dec.	33.2	33.4	0.52	0.51	2.7	3.4	66	65
Annual	54.0°	54.9°	21.22″	20.58″	18.1″	19.0″	69%	71%

[a] Current normals.
[b] Includes melted snow.
[c] Percentage of total possible.

degrees) was in 1904-05. And in Texas, during the winter of 1898-99, both Amarillo (29.5 degrees) and Fort Worth (39.3 degrees) registered all-time low winter means.

Although winters may be colder on the southern Plains in the near future, Dodge City's records suggest that Marches, on the average, may be a bit milder. This pattern of significant March warming, as I mentioned in the previous chapter, is one that may appear from the intermountain West eastward across the southern Plains all the way to the Appalachians.

Offsetting the March warming on the southern Plains will likely be noticeable cooling of the average monthly temperatures in July, August, and October. As in the northern Plains, this cooling may lead to some abnormally early frosts on the southern Plains—this despite the fact that little average temperature change is suggested for September. Dodge City's earliest freeze ever came on September 17, 1903. (The latest on record there was on May 27, 1907.) Other record early freezes were noted at Kansas City, Missouri, on September 30, 1895; and at Fort Worth, Texas, on October 22, 1898.

So, the older temperature data from the southern Plains support the notion that the next 20 or 30 years should offer colder winters; milder Marches; cooler Julys, Augusts, and Octobers; and the threat of unusually early frosts. Based on climate change alone, October-through-March heating bills might increase by an average of 4 or 5 percent, but June-through-August air-conditioning charges could drop by 10 percent or more.

Along with cooler weather, the coming decades should also bring somewhat damper weather. The Dodge City figures indicate about a 3-percent increase in yearly precipitation. Certainly there were some very wet years on the southern Plains around the turn of the century. Concordia, Kansas, had its wettest year on record in 1908, when 41.88 inches fell. Oklahoma City, Oklahoma, was soaked by 52.03 inches in the same year, its wettest ever.

There were some occasional dry years, too. Oklahoma City's driest came just 7 years before its wettest. Only 15.74 inches of water fell on Oklahoma City in 1901. The years 1893 and 1894 had large precipitation deficits across much of the High Plains region; and the phrase, "In God we trusted, in Kansas we busted," had become current as drought took its toll.[2]

Still, it was the damper weather that held sway during 1880–1909. Dodge City's wettest month was May 1889, when 12.82 inches of rain pelted down. The record for 24-hour rainfall was set in June 1899: 6.03 inches. In Texas, McKinney set a state record in May 1881 with a soaking 34.85 inches.

Older snowfall measurements from Dodge City suggest that annual snowfall totals in the coming decades will not average out significantly different than contemporary ones. There may be more snow, on the average, in February, and less in March, however. That would be consistent with the implied temperature changes for those months.

The plains south of Nebraska did get some noteworthy snow events 80 to 100 years ago. Dodge City's snowiest month and biggest storm came in February 1903. The total fall for the month was 27.7 inches, and 20.5 inches of that came in one storm. Amarillo's snowiest month was also February 1903. 28.7 inches of snow fell and led to a record depth of 16.5 inches.

Topeka was dumped on by more snow (27.1 inches) in February 1900 than in any other month on record. That same month also brought Topeka's single greatest snowstorm: 18.7 inches. Abilene, Texas, set a single storm record of 9.4 inches in February 1890.

All in all, the southern Plains should not only be cooler in coming decades but wetter as well. That is not to say that some drought years won't crop up from time to time, especially around the year 2000, as tentacles of the potential drought in the southwestern United States extend eastward (chapter 3). Overall snowfall totals may remain about the same—but watch out for Februarys.

Watch out for the 1990s, too. As I pointed out in chapter 3, the highest risk of major tornado activity may move into the plains of Texas, Oklahoma, and Kansas during that decade.

Notes

1. Ludlum, D. *Weather Record Book*, (Weatherwise, Inc., 1971), p. 27.
2. Ibid. p. 38.

CHAPTER 8

The Great Lakes

I HAVE HEARD it said that Chicago is not the greatest place to be in the winter. If that is the case, then the news that winters in Chicago and the rest of the Great Lakes area are probably going to be colder and snowier in coming decades is not going to be welcome.

Actually, average *annual* snowfall might decrease a bit in the Windy City in the near future. But Januarys, and especially Februarys, are going to see more of the white stuff piling up than has been the case over the past several decades.

Not to pick on Chicago, records from Detroit suggest the same thing: But there the increase in snowfall probably will not be confined to just January and February. A significant upward trend in yearly snowfall is possible in the Motor City.

The cooling around the Great Lakes should not be just a wintertime phenomenon over the next 20–30 years; all seasons should share in a drop in mean temperature.

The Western Great Lakes

Chicago records from the 1880–1909 period (table 15) imply that the next 2 or 3 decades around the western Great Lakes are going to be cooler and slightly drier. The greatest amount of cooling is suggested for the months April through August. However, the magnitude of the implied cooling in those months may be somewhat overstated. This is because the observations around the turn of the century were taken several miles closer to the shore of Lake Michigan than are contemporary readings. Lake Michigan, in the warmer season of the year, exerts a cooling influence—a "lake breeze" effect—on nearby land areas. Thus, in addition to any real differences between temperatures of 80 to 100 years ago and current normals, the effect of an observation site location change may be present. Still, the temperature trend should be

TABLE 15   Chicago, Illinois, Monthly Averages

	Temperature		Precipitation[b]		Snow		Sunshine[c]	
	1880-1909 (1980-2009)[?]	1941-1970[a]	1880-1909 (1980-2009)[?]	1941-1970[a]	1884-1909 (1980-2009)[?]	1941-1970[a]	1894-1903	1943-1976
Jan.	24.0°	24.3°	2.07″	1.85′	10.0″	9.4″	46%	44%
Feb.	25.0	27.4	2.58	1.59	11.7	8.1	53	47
Mar.	34.8	36.8	2.43	2.73	5.2	7.7	51	52
Apr.	46.3	49.9	2.70	3.75	0.7	1.0	62	54
May	56.7	60.0	3.47	3.41	0.1		63	61
June	66.4	70.5	3.39	3.95			68	67
July	72.2	74.7	3.47	4.09			71	70
Aug.	71.1	73.7	3.25	3.14			68	68
Sept.	65.1	65.9	3.07	3.00			63	64
Oct.	53.2	55.4	2.21	2.62	0.1	0.3	61	61
Nov.	39.6	40.4	2.48	2.20	2.1	2.7	42	41
Dec.	29.0	28.5	1.89	2.11	6.4	9.7	38	38
Annual	48.6°	50.6°	33.01″	34.44″	36.3″	38.9″	57%	57%

[a] Current normals.

[b] Includes melted snow.

[c] Percentage of total possible.

toward significant cooling, since the figures from Detroit (table 16) suggest the same tendency.

Chicago's coldest winter came in the 1880–1909 period; the winter of 1903-04 averaged just 18.3 degrees in the city. And a number of stations around the western Lakes set all-time-low-temperature records in the 1880s and 1890s. In Wisconsin in January 1888 residents of Green Bay (36 below) and Madison (29 below) shivered in a record-breaking cold wave. Even alcohol would have frozen in Duluth, Minnesota, in January 1885, when the mercury bottomed out at 42 below. Indianapolis registered its all-time minimum temperature in January 1884 with a reading of

TABLE 16   Detroit, Michigan, Monthly Averages

	Temperature		Precipitation [b]		Snow		Sunshine [e]	
	1880–1909 (1980–2009)?	1941–1970 [a]	1880–1909 (1980–2009)?	1941–1970 [a]	1880–1909 (1980–2009)?	1941–1970 [a]	1891–1903	1934–1965
Jan.	24.9°	25.5°	1.96″	1.93″	10.4″	8.1″ [c]	36%	32%
Feb.	24.7	26.9	2.42	1.80	10.1	7.2 [c]	42	43
Mar.	33.5	35.4	2.23	2.33	7.8	5.5	48	49
Apr.	46.2	48.1	2.29	3.08	2.3	1.1	53	52
May	57.9	58.4	3.24	3.43	0.2		57	59
June	67.5	69.1	3.91	3.04			66	65
July	72.1	73.3	3.36	2.99			70	70
Aug.	70.1	71.9	2.73	3.04			65	65
Sept.	64.1	64.5	2.52	2.30			64	61
Oct.	51.8	54.3	2.35	2.52			57	56
Nov.	39.3	41.1	2.55	2.31	2.6	2.5 [d]	36	35
Dec.	29.3	29.6	2.33	2.19	9.0	7.0 [d]	31	32
Annual	48.5°	49.9°	31.89″	30.96″	42.4″	31.4″	52%	54%

[a] Current normals.

[b] Includes melted snow.

[c] 1967 missing.

[d] 1966 missing.

[e] Percentage of total possible.

25 below. In Michigan, Marquette's coldest day ever (27 below) came in February 1888, and Muskegon's iciest reading (30 below) was recorded in February 1899.

The older weather records also indicate that despite the implied drift toward cooler weather, heat waves will still blister the region from time to time. Marquette's record high temperature of 108 degrees came in July 1901.

Chicago rain and snowfall figures from 1880 through 1909 suggest a slight decrease in average annual precipitation for 1980–2009, with the bulk of the decrease coming in April, June, and July. Certainly there were some dry years around the western Lakes near the turn of the century. Madison, Wisconsin's driest year ever was 1895, when just 13.63 inches of water fell. In 1901, Milwaukee, Wisconsin, had but 18.69 inches of precipitation, the least ever for that city.

But there were plenty of wet years, too. In Wisconsin, a state record was established in 1884, when 62.07 inches of precipitation soaked a town called Embarras. Madison's wettest year ever was 1881, when 52.93 inches of rain (and melted snow) fell. In Michigan Grand Rapids's soggiest year was 1883 (52.14 inches); Muskegon's wettest was 1905 (43.98 inches), as was Sault Sainte Marie's (42.96 inches).

So, while regions in the vicinity of the western Great Lakes may average slightly drier in coming decades, there should be plenty of wet years, too. That would be consistent with the pattern (wetter weather) that will probably prevail over most of the United States.

Despite the fact that annual snowfall totals around Chicago, and perhaps much of the western Lakes area, may decrease slightly over the next 20 or 30 years, snow shovels and snowplows may be even busier in January and February than current normals indicate. Contemporary average snowfall in Chicago for the months of January and February is 17.5 inches. During 1880–1909 it was almost 22 inches. And the mean temperature for February during that same time was almost 2½ degrees colder than it was during 1941–1970. Thus, February, in the near future, seems destined to become an even more wintery month than it has been recently.

Even though yearly snowfall amounts may diminish a bit over the next several decades, snowfall records from 1880 through 1909 suggest that that will not preclude some very snowy years from occasionally occurring. In Wisconsin Madison and Milwaukee both had their snowiest winter ever in 1885-86. In Madison almost 76 inches of snow fell, while just under 110 inches filtered down on Milwaukee. Green Bay established its maximum depth record in February 1893, when almost 2 feet of snow was measured on the ground. Indianapolis had its snowiest season in 1895-96: 46.8 inches. And residents of Marquette, Michigan, had to plow, shovel, and maybe even tunnel through, a record-breaking seasonal snowfall of almost 16 feet in 1890-91!

In summary, the western Great Lakes will most likely be colder and slightly drier over the next 20 to 30 years. Januarys, and particularly Februarys, will probably see more cold waves and snowstorms than has been the case over the past 20 or 30 years. Summers, too, should share in the cooler weather, maybe with a little less shower and thunderstorm activity.

By the way, Chicago may be the Windy City, but it is not the windiest city. Politics aside, that honor should go to Boston, Massachusetts. Boston has an average annual wind speed of 12.6 miles per hour, whereas Chicago checks in with a mere 10.4. Remember that next time you are chasing your hat through the Loop.

The Eastern Great Lakes

Climatological data from Detroit (table 16) suggest for the coming decades a trend which may already have started: a significant increase in average annual snowfall around the eastern Great Lakes. The mountainous snows in Buffalo, New York, in the winter of 1976-77 and the deep snow cover across Ohio in the winter of

1977-78 may have been harbingers. Those record-shattering events may not necessarily be surpassed in the next few decades, but they could be approached more frequently than in the recent past.

The 1880–1909 figures from Detroit imply an increase in yearly mean snowfall of almost a foot! The largest monthly increase is indicated for February. Detroit's snowiest month ever was February 1908, when 38.4 inches fell. In northern Michigan Alpena was besieged by record-breaking snow in the winter of 1886-87: 51.9 inches fell in January 1887, with the final total for that season reaching over 136 inches.

The absolute maximum snow depth at Alpena was attained in February 1904, when 33 inches were measured on the ground. Detroit's maximum depth of 26 inches was reached in March 1900.

In Ohio Akron's snowiest month on record was April 1901, when 26.6 inches' worth of late season flakes fell. In January 1893 Sandusky measured a total fall of 29.8 inches—a record for that city.

So, the message for the eastern Great Lakes states seems pretty clear: lots of snowy winters as the twenty-first century approaches.

As if the snow would not be enough, winters should be colder, too. Detroit's readings from 1880–1909 suggest that February will show the largest dip in mean temperature. As a matter of fact, on the average, it may nudge out January for coldest-month-of-the-year honors.

In Ohio Milligan set a state low temperature record in the great cold wave of February 1899, when the thermometer bottomed out at 39 below. Columbus and Dayton each established all-time minimum temperature marks in that same super Arctic outbreak. Columbus tagged 20 below, and Dayton plunged to 28 below.

Also, around the turn of the century, residents of both Detroit and Sault Sainte Marie, Michigan, endured the coldest winter on record in those cities. For both locations it was the winter of 1903-04, the same as Chicago's coldest. Detroit's average temperature that winter was 18.7 degrees, while at Sault Sainte Marie it was just 8.4 degrees.

The cooling around the eastern Lakes should be apparent in all seasons of the year, with summers, too, frequently being significantly cooler than contemporary normals indicate. However, October may exhibit the largest average cooling for any one individual month.

Despite the suggested overall cooling, there should not be a complete absence of hot spells. Dayton, Ohio, sweltered in 108-degree heat in July 1901. It has never been hotter there.

The annual precipitation figures from 80 to 100 years ago suggest that the eastern Lakes will be a bit damper in the near future. Aprils and Augusts could be slightly drier, according to Detroit's data, but other months should be wetter or show little change from current averages.

Individual years from 1880 through 1909 were actually quite variable when it came to precipitation. Adrian, Michigan, set a state record for greatest annual precipitation in 1881 with 64.01 inches. And in Ohio Akron, Cincinnati, and Columbus all had record wet years in the late 1800s. But Detroit, Toledo, and again Cincinnati all had their driest year ever about the same time (between 1889 and 1901).

The final word for the eastern Great Lakes, then, is that on the average the immediate future is going to be colder, snowier, and probably a shade wetter, although any one year could be quite dry.

In addition to the myriad of other economic inflationary pressures in coming decades, residents of all the Great Lakes states are going to have to contend with those brought on by colder temperatures, too. In terms of heating fuel bills, that means a weather-sponsored 7- or 8-percent average increase on top of everything else.

There is some bright news, though. The cooler summers should result in less use of air conditioners, and hence, less use of costly electricity.

Who knows, maybe fans will come back in vogue.

CHAPTER 9

Mid-America

RECORDS FROM Saint Louis, Missouri, and Little Rock, Arkansas, represent the region I define as Mid-America. Specifically, I am considering eastern Iowa, most of Missouri (except for the extreme west), Arkansas, most of Illinois (except for the northeastern corner), southern Indiana, and the western parts of Kentucky and Tennessee as being Mid-America.

If you live in that area and you like the current climate, good. The 1880–1909 data from Saint Louis (table 17) and Little Rock (table 18) imply that things aren't going to change much over the next 20 or 30 years. On an annual basis, temperatures should not be noticeably different from current normals, and precipitation may increase only very slightly.

Individual months may display some important differences, though. The numbers from both Saint Louis and Little Rock suggest that Februarys will probably average a little colder in the near future, and that the months from November through January will be just a bit milder. Octobers may be slightly cooler on the average, but perhaps not significantly so. And most Marches will probably be somewhat milder than those of 1941–1970.

The Little Rock records suggest that summers will average out

TABLE 17   Saint Louis, Missouri, Monthly Averages

	Temperature		Precipitation [b]		Snow		Sunshine [c]	
	1880–1909 (1980–2009)?	1941–1970 [a]	1880–1909 (1980–2009)?	1941–1970 [a]	1884–1909 (1980–2009)?	1941–1970 [a]	1891–1903	1960–1976
Jan.	31.7°	31.3°	2.46"	1.85"	5.4"	4.0"	53%	53%
Feb.	33.6	35.1	3.04	2.06	5.1	4.0	51	53
Mar.	44.1	43.3	3.49	3.03	3.5	4.9	52	56
Apr.	56.6	56.5	3.46	3.92	0.8	0.1	58	58
May	66.2	65.8	4.56	3.86			63	62
June	75.0	74.9	4.18	4.42			67	69
July	78.9	78.6	3.32	3.69			70	71
Aug.	77.7	77.2	2.31	2.87			71	67
Sept.	70.9	69.6	2.96	2.89			71	64
Oct.	58.8	59.1	2.41	2.79			70	63
Nov.	45.6	45.0	3.08	2.47	0.7	1.1	52	50
Dec.	35.7	34.6	2.13	2.04	3.3	3.0	49	43
Annual	56.2°	55.9°	37.40"	35.89"	18.8"	17.1"	60%	59%

[a] Current normals.
[b] Includes melted snow.
[c] Percentage of total possible.

cooler in the coming decades, but the Saint Louis figures do not support that. That lack of support may be artificially induced, however. Contemporary observations at Saint Louis are taken at Saint Louis International Airport, outside the city center. Thus, the 1941–1970 means are probably biased when compared with the older city-based averages. That is, any natural warming that may have taken place relative to 1880–1909 would be masked by the cooler contemporary averages, which are a function of a geographical change in the observation site. Therefore, it may be that Saint Louis, also, can expect to experience a greater frequency of more moderate summers in the near future; not that the summers will be cool by any stretch of the imagination, they just will not be quite so hot.

TABLE 18   Little Rock, Arkansas, Monthly Averages

	Temperature		Precipitation b		Snow		Sunshine d	
	1880–1909 (1980–2009)?	1941–1970 a	1880–1909 (1980–2009)?	1941–1970 c	1880–1909 (1980–2009)?	1941–1970 c	1894–1903	1945–1976
Jan.	41.7°	41.1°	4.86″	4.24″	2.4″	2.4″	48%	46%
Feb.	44.2	44.5	4.24	4.42	1.2	1.5	53	54
Mar.	53.5	51.7	4.74	4.93	0.6	0.6	54	57
Apr.	62.7	63.0	4.36	5.25			65	61
May	70.1	70.8	5.34	5.30			63	68
June	77.5	78.8	3.94	3.50			73	73
July	80.6	81.8	3.82	3.38			72	71
Aug.	79.9	81.0	3.29	3.01			81	73
Sept.	74.0	74.0	3.29	3.55			74	68
Oct.	63.3	63.7	2.42	2.99			71	69
Nov.	51.9	51.5	4.70	3.86	0.2	0.1	56	56
Dec.	44.1	43.2	4.07	4.09	1.1	1.0	53	48
Annual	62.0°	62.1°	49.07″	48.52′	5.5″	5.6′	64%	63%

a Actual, unadjusted means.
b Includes melted snow.
c Current normals.
d Percentage of total possible.

Though annual precipitation totals may be a bit higher in many of the coming years, the period August through October may actually average a bit drier and sunnier than current normals indicate. At least that is the implication of the 1880–1909 figures from Saint Louis and Little Rock. The numbers also suggest that Aprils, too, will show significantly less rainfall in the near future. The most important increases in monthly precipitation means will likely come in the months of January and November.

Snowfall measurements from 80 to 100 years ago imply a small increase in seasonal totals over the next few decades at Saint Louis, but little change at Little Rock.

Although winters as a whole probably will not be much tougher over the next 25 years than they have been over the past several decades, the month of February could bring some numbing cold waves. February 1905 saw an Arctic outbreak roar southward out of Canada and set state minimum temperature records in both Missouri and Arkansas. The mercury tumbled to 40 below in Warsaw, Missouri, and slid to 29 below in Pond, Arkansas. Little Rock's thermometer, at a record 13 below, was a victim of the super cold wave of February 1899. In that same historic polar blast all-time low marks were reached in Columbia, Missouri (26 below); Springfield, Missouri (29 below); and Fort Smith, Arkansas (15 below). Other record minima were recorded in the February 1905 cold snap when Burlington, Iowa, hit 27 below, and Springfield, Illinois, 23 below.

Additional low marks from around the turn of the century were January products. Dubuque, Iowa, tagged 32 below in 1887. And in Illinois both Peoria and Moline rang up identical readings of 27 below in 1884.

The suggestion that yearly snowfall totals will increase only slightly or not at all seems to complement the implication that winter temperature averages will not change very much. No annual snowfall records were established in the Mid-America region during 1880–1909, but there were some notably snowy individual months.

An Illinois state record for monthly snowfall was established at Astoria in February 1900, when almost 4 feet (47 inches) fell. The same month also produced records at Peoria (26.5 inches) and Springfield (24.4 inches), Illinois. A 32-inch fall, a record, blanketed Urbana, Illinois, in March 1906.

A single storm in March 1892 dumped enough frozen precipitation on Memphis and Nashville, Tennessee, to establish monthly snowfall marks in both of those cities: 18.5 inches at Memphis and 21.5 inches at Nashville.

A number of records for maximum snow depth were also written into the archives around the turn of the century. As a result of its heaviest 24-hour snow ever, Saint Louis measured 20.4

inches on the ground in March of 1890. Little Rock's deepest snow, 10 inches, piled up in January 1893. That snow, too, was the product of a record-breaking 24-hour fall: 13 inches.

Later that same winter (1892-93) Dubuque, Iowa, reported its deepest snow ever: 27 inches in February. Peoria, Illinois, checked in with 18 inches on the ground—a record—in February 1900. As a consequence of the March 1892 storm at Memphis and Nashville, both of those cities established all-time maximum depth records. In Memphis the snow was 18 inches deep, in Nashville, 17 inches.

Annual precipitation figures from 80 to 100 years ago suggest that superimposed on a slight overall average increase in the near future will be rather significant variability. This variability may operate on a time scale of about a decade or so. For example, between 1880 and 1890 several all-time records for yearly wetness were set. A state mark for sogginess was established in Indiana when Marengo, in the southern part of the state, was inundated by 97.38 inches of water in 1890. Little Rock's wettest year ever (71.20 inches) was 1884. Springfield, Illinois, checked in with a record 58.21 inches in 1882. And the mildew did well in Nashville in 1880, when 67.24 inches of precipitation poured down.

In the ensuing 15 years after 1890, several records for annual dryness were tallied. In parts of Missouri and Illinois the year 1901 brought drought conditions. Columbia, Missouri, with 21.25 inches, had its driest 12 months on record. And Moline, Illinois, could manage only 17.33 inches. Dubuque, Iowa, measured just 19.39 inches of precipitation in 1894, the most parched year on record there.

Thus, although climatological means may not change very much in Mid-America over the next several decades, individual years or months could show some important fluctuations. Not just in precipitation totals either. For instance, Februarys could produce somewhat more frequent and severe cold waves than in the recent past. And although winters as a whole might not set any snowfall records, there could be some abnormally snowy months, often the result of just a single monstrous snowstorm. Drought—as mentioned—may haunt the region from time to time, but persistent, severe drought does not seem likely.

All in all, most of Mid-America probably won't experience any more weather problems from 1980 through 2009 than it did from 1941 through 1970. I suppose the fact that late summer through early autumn (August through October) could average our drier and sunnier may even be a meteorological bonus.

There is an important negative aspect to the outlook for the next 20 years, however. It is something that could be especially critical to the residents of Arkansas. As I discussed in chapter 3, the maximum killer tornado threat will probably shift into Arkansas by the late 1980s and remain in that general region through the early 1990s. That outlook is not meant to alarm. It is meant to encourage alertness and responsiveness to warnings issued by the National Weather Service and local agencies.

Mid-America has always had to deal with tornadoes. That will not change in the immediate future.

CHAPTER 10

The South

I GUESS if I had my choice of anywhere in the country to spend springs, I would opt for the Deep South. That is a land of sunshine and warmth, and where magnolias bloom early. As a matter of fact, I would be willing to trade three Marches in Massachusetts for a single April in Alabama. That trade may become even more attractive in the near future. The 1880-1909 weather records imply that over the next few decades springs in New England—where I live—are going to arrive later (be colder), whereas springs in the Old South, by virtue of milder Marches, are going to appear earlier.

Compared with the means of 1941-1970, average annual temperatures throughout most of the South probably will not change very much during the next quarter century. But all the records from the region agree on one point: The implication that most Marches will be a bit milder. Offsetting that is the unanimous suggestion that Octobers may be slightly cooler—which could be welcomed after long, hot summers.

The indications for average yearly precipitation totals are variable, ranging from a moderate increase to a slight decrease. I will discuss this in more detail later in the chapter.

Perhaps even more important than changes in precipitation and temperature will be some changes in the potential threats of the more spectacular-type weather phenomena: tornadoes, hurricanes, and snowstorms. I suspect that meteorologists in the Old South will earn their keep over the next few decades.

Tornadoes

As I pointed out in chapter 3, it is the southern United States that may experience the greatest frequency of death-dealing tornadoes in the 1980s. If the cycle that T. Theodore Fujita and his coworkers discovered continues, then the maximum threat of killer tornadoes will move from the Carolinas across Georgia, Alabama, and Mississippi into Louisiana and Arkansas during the 1980s. This is not saying that the South will be hit by *more* twisters than has recently been the case. Rather it is suggesting that during the decade of the 1980s more of the *major* tornado outbreaks will ravage areas of the Deep South than they will other regions of the country.

That outlook is not designed to leave you cowering in your basement day after day during tornado season. But I hope that if I mention the possibility of major twisters in the South, you—if you live there—will give some thought to the problem. At a minimum I suspect you will want to know how your community issues tornado warnings. Is a siren sounded? Are the warnings broadcast on local radio stations? Is there a NOAA Weather Radio station within range of you?

Assuming you know the answers to the preceding questions, do you know how to react when a warning is actually sounded? Your reaction can be different depending upon your circumstances. You might be caught in the open; in a car; in a mobile home; or in a large office building. If you are unsure of what a warning means or how to respond to it, contact officials from your local Civil Defense office or the National Weather Service (NWS). Surviving a tornado threat is not a function of becoming paranoid about twisters. It is a function of staying alert for warnings and knowing how to react to those warnings.

The cheapest tornado insurance you can invest in is probably one of the small weather radios that offers an alerting feature; by sounding a tone or flashing a light, the radio lets you know that a warning has been issued by the NWS.

Hurricanes

As well as being threatened by major tornado outbreaks over the next 10 or 20 years, the South—at least that portion that borders the Atlantic Ocean—may have to contend with a greater frequency of major hurricane threats. Compared with coastal regions north of Cape Hatteras, North Carolina, and with those along the Gulf of Mexico, the Atlantic shore areas between Hatteras and Key West, Florida, will probably experience more brushes with severe hurricanes during the next several decades. The comparative threat along the Gulf Coast may remain about the same or perhaps diminish a bit. I discussed the reasons in chapter 3.

Still, it takes only one storm to make statistical threats personally meaningless. The three deadliest hurricanes on record in the United States struck during the 1880-1909 period, and two of them ripped into Gulf Coast states.

Over 6,000 people were killed, mostly by drowning, in Galveston, Texas, in September 1900, when a hurricane-induced storm tide rolled over the already inundated city. Hundreds of homes and buildings were washed away in the worst weather disaster in United States history.

In October 1893 a savage hurricane made landfall in Louisiana between New Orleans and Port Eads. Damage along the Louisiana and Mississippi coasts was reported as phenomenal. The bayou country was overwhelmed by wind-driven Gulf waters, and 1,800 to 2,000 people perished.

In August of that same year (1893) a fierce tropical invader swept inland between Savannah, Georgia, and Charleston, South Carolina. A tremendous storm wave submerged all of the coastal islands. Between 1,000 and 2,000 lives were lost.

Will the next few decades bring a repeat of such deadly storms? No, because the great loss of life around the turn of the century was more a function of inadequate warnings and lack of preparedness than it was of hurricane severity. That is not to say the storms were not awesome. They were. But even the most awesome storms can now be accurately tracked and forecast by the NWS. Precise warnings can be given and thousands of lives saved.

Hurricane Camille, the greatest storm of any kind ever to strike the United States, killed 256 people in August 1969. That is about one twenty-fifth the death toll of the great Galveston disaster. The point is, no matter how fearsome the storm, the modern warning system of the NWS can save lives.

It can save lives, however, only if the warnings are heeded.

Snowstorms

The record-breaking snowstorm that swept across parts of the Deep South in February 1973 may have been the first of several such events that will have return engagements over the next few decades. A storm on the ninth and tenth of February, 1973, laid down a blanket of white in excess of a foot deep across south-central Georgia, central South Carolina, and southeastern North Carolina. The period 1880–1909 saw several similar occurrences in other parts of the South, a suggestion that 1980–2009 will also play host to more frequent southern snowstorms.

A memorable snowstorm along the upper Texas and central Gulf Coast in February 1895 has been termed by weather historian David Ludlum as "the greatest deep-snow anomaly in the recorded climate history of the U.S."[1] The storm's effects were felt as far south as Brownsville, Texas, on the Mexican border, where the official snowfall measurement was 3 inches. But local newspaper accounts mentioned depths of up to 6 inches. Corpus Christi, Texas, picked up 4.3 inches from the storm, but this was surpassed by a 5-inch fall in 1897.

The great 1895 storm went on to dump all-time record snowfalls from just north of Corpus Christi northeastward to Louisiana and then eastward into Georgia and Florida. Residents of Galveston, Texas, watched in amazement as 15.4 inches of snow piled up, while a short distance inland an astounding 20 inches came down on Houston. A contemporary 20-inch fall can completely paralyze a city such as Boston or New York for several days. Can you imagine the effect of such a storm on the Sunbelt city of Houston?

From the Houston-Galveston area the 1895 storm really opened up on Louisiana. Lake Charles collected 22 inches, and Rayne (near Lafayette) established a state record with a total accumulation of 2 feet! Shades of Buffalo. New Orleans also tallied its deepest snow ever with a fall of 8.2 inches. The storm continued to be a prolific snow maker as it moved through Alabama, Georgia, and Florida. Mobile, Alabama, set a record with a 6-inch fall. Pensacola, Florida, measured a record 3 inches, while Thomasville, in southern Georgia, set an all-time mark with 4 inches.

Thanks mostly to that single storm, Atlanta, Georgia, established monthly (11.6 inches) and seasonal (13.2 inches) snowfall records; Diamond, in the mountains of north-central Georgia, did the same with a monthly total of 26.5 inches and a seasonal tally of 39 inches. Both of those figures are Georgia state records.

Another significant snowstorm occurred farther east in February 1899. Jacksonville, Florida, collected almost 2 inches of snow from that storm. That amount has never been equaled there. Snow fell as far south as Tampa, Florida, where the official measurement was 0.1 inch. A similar amount had fallen in

January 1886. And it was not until January 1977 that enough snow to measure again fell on Tampa.

The 1899 storm dropped 2 inches of snow on Savannah, Georgia, and almost 4 inches on Charleston, South Carolina. Inland in South Carolina the storm contributed enough snow to Columbia to make 1898-99 the snowiest season (16.7 inches) on record there.

A great snowstorm blitzed the entire state of Alabama in December 1886. Average depths ranged from 12 to 16 inches in the central part of the state, to 17 to 20 inches in the north. Montgomery, with 11 inches, recorded its greatest snowstorm ever. After burying Alabama the storm moved into northern Georgia, where people around Rome were stunned as over 2 feet (25 inches) of snow piled up.

Other storms of lesser extent occurred early in the twentieth century. Jackson, Mississippi, had its deepest snow on record in January 1904: 11.4 inches. And Spartanburg, South Carolina, had its most memorable snow in February 1902 with 15 inches. A record for a season was established that same winter with a total snowfall of 17.5 inches.

The beginning of the last 80-year long-term solar cycle—the cycle that began 180 years ago and which should be similar to the one projected to start in 1980—was about 1800. Then, too, unusually deep snows plagued parts of the South. In January 1800 a survey team along the Georgia-Florida border north of Tallahassee measured a 5-inch fall. Depths of equal or greater magnitude have not since been recorded in that area. January 1800 also saw a storm blanket Savannah, Georgia, with 18 inches of snow, a depth that has not been approached there in the past 180 years.

In that same winter (1799-1800) Charleston, South Carolina, was bombed by four significant snows that totaled up to 14 inches. No winter since then has seen that mark threatened in Charleston.

But perhaps the respite from the threat of deep southern snows is about over now, if indeed it has not already ended.

Precipitation

The indications for precipitation trends over the South during the next few decades are diverse. The records from 1880–1909 at Galveston, Texas, and Augusta, Georgia, both suggest an upward trend in annual rainfall, whereas the data from Mobile, Alabama, and Jacksonville, Florida, imply the opposite. Galveston's record is probably the best from which to draw conclusions because the observation site there has been in virtually the same location since 1871.

TABLE 19   Galveston, Texas, Monthly Averages

	Temperature		Precipitation [b]		Snow [c]		Sunshine [d]	
	1880–1909 (1980–2009)?	1941–1970 [a]	1880–1909 (1980–2009)?	1941–1970 [a]	1880–1909 (1980–2009)?	1941–1970 [a]	1890–1903	1891–1976
Jan.	53.9°	53.9°	3.44″	3.02″			47%	48%
Feb.	55.7	56.2	3.16	2.67			44	51
Mar.	62.6	61.0	2.93	2.60			47	55
Apr.	69.3	69.2	2.94	2.63			58	61
May	75.5	75.9	3.15	3.16			66	68
June	81.2	81.3	4.40	4.05			72	75
July	83.1	83.2	3.95	4.41			70	72
Aug.	83.0	83.3	4.39	4.40			67	71
Sept.	79.8	80.0	5.82	5.60			68	68
Oct.	72.6	73.1	4.07	2.83			73	72
Nov.	63.6	63.5	3.48	3.16			59	60
Dec.	56.7	57.1	3.22	3.67			51	49
Annual	69.7°	69.8°	44.95″	42.20″			60%	63%

[a] Current normals.
[b] Includes melted snow.
[c] Not significant.
[d] Percentage of total possible.

The figures from Galveston (table 19) indicate that about a 6½-percent increase in yearly rainfall is possible over the next 20 to 30 years. But superimposed on that indication is the suggestion of great variability. Consider the following statistics from around the turn of the century: Galveston's wettest year ever was 1900 (the year of the great hurricane), when 78.39 inches of rain pounded down, but there was also a 9-year stretch (1891–1899) when every year had below-normal precipitation; the wettest month on record was September 1885 with 26.01 inches of precipitation, but the driest month in history occurred within 17 years of that date: August 1902 had no rain whatsoever. Six records for maximum monthly precipitation were established between 1880 and 1909,

TABLE 20   Mobile, Alabama, Monthly Averages

	Temperature		Precipitation b		Snow c		Sunshine d	
	1880–1909 (1980–2009)?	1941–1970 a	1880–1909 (1980–2009)?	1941–1970 a	1880–1909 (1980–2009)?	1941–1970 a	1880–1909	1941–1970
Jan.	50.9°	51.2°	4.97″	4.71″				
Feb.	53.9	54.0	5.79	4.76				
Mar.	60.2	59.4	6.85	7.07				
Apr.	66.8	67.9	4.33	5.59				
May	73.7	74.8	4.48	4.52				
June	79.8	80.3	5.78	6.09				
July	81.4	81.6	6.61	8.86				
Aug.	81.1	81.5	6.82	6.93				
Sept.	77.9	77.5	5.17	6.59				
Oct.	68.3	68.9	3.09	2.55				
Nov.	58.9	58.5	3.23	3.39				
Dec.	52.3	52.9	4.71	5.92				
Annual	67.1°	67.4°	61.83″	66.98″				

a Current normals.
b Includes melted snow.
c Not significant.
d Figures not available.

but seven all-time marks for minimum monthly precipitation were tallied in the same period!

Thus, indications for much of southeastern Texas and western Louisiana seem to point toward slightly more rainfall in the near future, but with a great deal of year-to-year and month-to-month variation. That is, some droughty periods are likely in addition to the damp sieges.

Eastward along the Gulf Coast, the data from Mobile (table 20) suggest a decrease in average annual rainfall for that area, which does not jibe too well with the fact that Mobile's wettest month in history and wettest year ever both occurred in the years 1880–1909. The soggiest month was June 1900 (26.67 inches) and the rainiest year, 1881 (92.32 inches). The problem with Mobile's data is that they have been compiled at different locations, which may have significantly altered annual precipitation totals.

Observations around the turn of the century were taken in the city, adjacent to Mobile Bay. But contemporary observations are made at the airport, about 11 miles west of and inland from the city. That is just far enough away from the stabilizing influence (relatively speaking) of Mobile Bay to be meteorologically conducive to more thunderstorm buildups in the summer. If you will look at Mobile's record, you will see that, generally speaking, the largest differences between current monthly rainfall averages and the older normals occur in the warmer months of the year, April through September. Those are the months when the most thunderstorms rumble across the Mobile area.

If the data from the colder season of the year, October through March, are examined, very little difference between current means and turn-of-the-century normals is apparent. From October through March the precipitation regime is usually controlled by large-scale weather phenomena—the storms, or *lows*, the TV weatherman points out to you each evening—rather than by local thunderstorms. Thus, probably the best assumption to make for Mobile, and much of Louisiana, Alabama, and Mississippi, is that

TABLE 21   Augusta, Georgia, Monthly Averages

	Temperature		Precipitation [b]		Snow		Sunshine [d]	
	1880–1909 (1980–2009)?	1941–1970 [a]	1880–1909 (1980–2009)?	1941–1970 [c]	1880–1909 (1980–2009)?	1941–1970 [c]	1880–1909	1941–1970
Jan.	46.7°	46.5°	3.95″	3.44″	0.2″	0.1″		
Feb.	49.3	48.9	4.53	3.75	0.4	0.2		
Mar.	56.5	55.1	4.48	4.67				
Apr.	63.6	64.3	3.27	3.37				
May	72.5	72.2	3.19	3.38				
June	79.0	78.6	4.71	3.66				
July	81.3	80.9	5.09	5.09				
Aug.	80.2	80.1	5.71	4.21				
Sept.	75.6	74.8	3.36	3.26				
Oct.	64.5	64.7	2.34	2.17				
Nov.	54.9	54.3	2.33	2.21				
Dec.	47.8	46.9	3.24	3.42	0.3	0.1		
Annual	64.3°	63.9°	46.20″	42.63″	0.9″	0.4″		

[a] Actual, unadjusted means.
[b] Includes melted snow.
[c] Current normals.
[d] Figures not available.

average yearly rainfall will change very little in the near future. Although some very wet years, such as 1881, may crop up from time to time, so may some very dry ones. New Orleans, Louisiana, had its most parched year ever (31.07 inches) in 1899, and Birmingham, Alabama's driest (34.32 inches) came in 1904.

The implication from the Augusta record (table 21) is that there will be a healthy 8½-percent increase in mean yearly precipitation over the next 20 to 30 years. This is an increase of the order of magnitude that, as you will see in the following chapter, may extend northward through the Eastern Seaboard states as far as New York City. Supporting the indication of increased precipitation totals in the northern Georgia–South Carolina area are a couple of records from the 1880–1909 era. Greenville, South Carolina, tallied its wettest year ever (72.50 inches) in 1908, and Atlanta's soggiest month in history (15.82 inches) was January 1883.

The numbers from Jacksonville (table 22) suggest a slight decrease in precipitation for most years in the coming decades. But, as happened in Mobile, the switch of the Jacksonville observation site may have affected precipitation totals. Older records were kept in downtown Jacksonville, whereas observations during 1941–1970 were made at the airport north of the city. As you can see in table 22, significant increases in average precipitation for

TABLE 22 Jacksonville, Florida, Monthly Averages

	Temperature		Precipitation b		Snow d		Sunshine e	
	1880–1909 (1980–2009)ᵖ	1941–1970 ᵃ	1880–1909 (1980–2009)ᵖ	1941–1970 ᶜ	1880–1909 (1980–2009)ᵖ	1941–1970 ᶜ	1898–1903	1951–1976
Jan.	55.3°	55.6°	3.07′	2.78′			50%	57%
Feb.	57.8	57.3	3.03	3.58			53	61
Mar.	63.2	62.7	3.26	3.56			65	66
Apr.	68.6	69.8	2.62	3.07			76	71
May	75.1	76.2	4.34	3.22			73	69
June	80.0	80.6	5.28	6.27			66	62
July	82.0	82.4	6.87	7.35			61	60
Aug.	81.7	82.1	5.90	7.89			64	58
Sept.	78.3	79.0	8.13	7.83			53	54
Oct.	70.5	71.1	4.59	4.54			52	57
Nov.	62.5	62.2	2.14	1.79			52	61
Dec.	56.2	56.3	2.90	2.59			50	56
Annual	69.3°	69.6°	52.13′	54.47′			60%	61%

a Actual, unadjusted means.
b Includes melted snow.
c Current normals.
d Not significant.
e Percentage of total possible.

the months of June through August occurred in 1941–1970 relative to 1880–1909.

This again suggests that the observation location between 1941 and 1970 might have been situated such that it was more susceptible to local shower and thunderstorm activity than was the older city site. But the physical reason for this is not so clear as it was in Mobile's case. Nonetheless, probably the safest conclusion to draw for the Jacksonville area, and for much of the rest of northern Florida and southern Georgia, is that average annual precipitation totals will decrease very slightly in the near future.

Temperature

Average yearly temperatures in the South during 1980–2009 will probably not show much change from 1941–1970 normals. But means in certain individual months could. For instance, the implications from the 1880–1909 figures are that Marches will average a bit milder and Octobers slightly cooler in the near future.

And although winter temperatures, when averaged over a 30-year period, may not show much change from contemporary normals, some notably chilly winters could occasionally pop up. Just so southerners do not get too smug about the prospects of never having to contend with cold winters over the next few decades, let us consider some statistics from the 1880–1909 era.

Several stations in southeastern Texas and Louisiana had their coldest winter on record during that period. Brownsville, Texas, checked in with its coldest winter in history (44.6 degrees) in 1898-99. Inhabitants of San Antonio shivered in record cold through the winter of 1904-05 when the average temperature for December through February was 48.9 degrees. The winter of 1885-86 was New Orleans's chilliest with a mean of 50.9 degrees.

Galveston's coldest month ever (42.6 degrees) was February 1895, the same month of the great snow. Brownsville checked in with an all-time-low monthly mean in January 1881: 40.9 degrees. In Florida the Keys had their coolest month on record in January 1905, when the mean at Key West was just 63.4 degrees.

Record minimum temperatures were established at numerous locations throughout the South near the turn of the century. And they were all the product of just one super-cold Arctic outbreak: the colossal cold wave of February 1899. That awesome cold snap started thermometers diving to record lows in the state of Nebraska and did not relent in that task until it reached the Atlantic Ocean.

The great onslaught of frigid air swept through Dixie. State records for coldness were set in Louisiana and Florida: In Minden, Louisiana, the mercury sank to 16 below, while Tallahassee, Florida, registered a remarkable 2 below. The list of absolute minimum temperatures established in other locations during that great icy outbreak is long:

Brownsville, Texas	12 above
Corpus Christi, Texas	11 above
Galveston, Texas	8 above
Baton Rouge, Louisiana	2 above
New Orleans, Louisiana	7 above
Shreveport, Louisiana	5 below
Meridian, Mississippi	6 below
Vicksburg, Mississippi	1 below
Birmingham, Alabama	10 below
Mobile, Alabama	1 below
Montgomery, Alabama	5 below
Athens, Georgia	3 below
Atlanta, Georgia	9 below
Augusta, Georgia	3 above
Rome, Georgia	7 below
Savannah, Georgia	8 above

Jacksonville, Florida	10 above
Pensacola, Florida	7 above
Charleston, South Carolina	7 above
Columbia, South Carolina	2 below

Besides the historic minimum of 8 degrees in Galveston in February 1899, record lows for the other two winter months were also tallied in that city in the late 1800s. The December mark of 18 degrees was established in 1880, and the January nadir of 11 degrees was reached in 1886.

Despite the fact that winters may not average any colder in coming decades, Florida citrus crops may have to contend with severely damaging freezes more often. Peter Leavitt, a certified consulting meteorologist with Weather Services Corporation, analyzed winter minimum temperatures in the Florida citrus belt back through the season of 1880-81.[2] In the 30-year period starting with the winter of 1880-81 there were nine freezes that produced at least moderate-to-severe damage to citrus crops. In the 30-year period beginning with the winter of 1941-42 there were only four such freezes—all of them, by the way, occurring since 1957. Thus, the trend toward a greater frequency of dangerous Florida freezes, as suggested by the 1880–1909 data, may already have begun.

The figures from 1880–1909 seem to suggest that citrus growers can expect a freeze producing at least moderate-to-severe crop damage in 3 out of every 10 years in the near future. Freezes causing damage of at least a moderate (as opposed to a moderate-to-severe) category would come along more often than that, of course. *

*Leavitt's study indicated that such freezes tend to cluster into "regimes." That is, instead of icing up oranges and grapefruits in single, randomly distributed winters, the freezes tend to concentrate into spells lasting anywhere from 3 or 4 years to 15 or more. For instance, nine significant freezes might occur in 16 years, twelve in 14, and so on.

Even though record cold may be more likely to occur than record heat in the South in the near future, you should not count out the possibility of some dandy heat waves, too. The figures from around the turn of the century indicate that some blistering readings were observed in August 1909. Houston, Texas (108 degrees), San Antonio, Texas (107 degrees), Baton Rouge, Louisiana (110 degrees), and Shreveport, Louisiana (110 degrees), all set maximum temperature records that month. Montgomery, Alabama, with 107 degrees established a high mark in July 1881.

In summary, the South, although experiencing little change in overall temperature and precipitation averages during the next several decades, may have to contend with some rather extreme individual weather events. At the least, some record-breaking cold waves might be expected. But do not discount the threat of hot spells and droughts from time to time, either.

Big storms may become an important part of southern weather lore over the next 20 to 30 years, too. The older climatic data suggest that there will be a greater risk, than in the recent past, of traffic-snarling snowstorms and deadly tornado swarms. That is not to paint a picture of gloom and doom for the South. The countryside is not going to turn into an Arctic tundra or be swept bare by twisters. You will be as safe living in the South as anyplace in the country. All that is required is that you pay a modicum of attention to the weather warning services available to you.

Those magnolias will be around for a long time yet.

Notes

1. Ludlum, D. *Weather Record Book*, (Weatherwise, Inc., 1971), p. 15.
2. Leavitt, P. "On the Inter-Seasonal Relationships Among Minimum Temperatures and Damaging Freezes in the Central Florida Citrus Belt," *National Weather Digest* 3 (1978): 25; and personal communication with Peter Leavitt, February 1978.

CHAPTER 11

The East

THE EASTERN UNITED STATES will be colder and, over most of the region, snowier and wetter in coming decades. At least that is the implication from the 1880–1909 records. On an annual scale, the greatest amount of cooling, and the largest increase of rainfall and snowfall, should be centered around the Washington, D.C., area. As a matter of fact, the Washington records suggest more cooling in the near future than do the figures from any of the other twenty-three cities I studied.

It is perhaps poetic justice that winters in the near future may be as much, if not more, colder and snowier around our nation's capitol as anywhere in the country. Such a trend would manifest the significance of the 180-year cooling cycle in a place where climatic cycles are apparently not appreciated. In June 1974 Reid Bryson of the University of Wisconsin and Stephen Schneider of the National Center for Atmospheric Research were in Washington urging an audience of White House policymakers to establish a federal program to build up food reserves. Such reserves, Bryson and Schneider argued, would act as a hedge against such weather vagaries as the 20- to 22-year drought cycle. They were interrupted in their presentation by a sarcastic voice from

the back of the meeting room. "Around here," said the
voice—referring to the terms of office of congressmen, president,
and senators—"the only cycles that count are the two-, four-, and
six-year cycles."[1]

Northern New England and Upstate New York

Figures from Burlington, Vermont (table 23), suggest that much
of northern New England and upstate New York will average
around a degree and a half cooler over the next few decades.[2] The
maximum amount of cooling is implied for the early spring
(March and April) and late autumn (October and November).

TABLE 23   Burlington, Vermont, Monthly Averages

	Temperature		Precipitation [c]		Snow		Sunshine [d]	
	1884–1909 (1980–2009)? [a]	1941–1970 [b]	1880–1909 (1980–2009)?	1941–1970 [b]	1884–1909 (1980–2009)?	1941–1970 [b]	1880–1909 [e]	1944–1976
Jan.	16.3°	16.8°	1.70″	1.74″	16.3′	17.0′		42%
Feb.	16.5	18.6	1.55	1.68	15.0	16.9		48
Mar.	26.5	29.1	1.98	1.93	13.6	11.7		52
Apr.	40.9	43.0	1.91	2.62	3.1	2.4		51
May	54.5	54.8	3.06	3.01	0.1	0.2		55
June	63.6	65.2	3.36	3.46				60
July	67.9	69.8	3.84	3.54				65
Aug.	65.2	67.4	3.65	3.72				62
Sept.	59.8	59.3	3.57	3.05				55
Oct.	46.2	48.8	2.76	2.74	0.2	0.2		50
Nov.	34.4	37.0	2.57	2.86	7.8	6.2		30
Dec.	21.9	22.6	1.84	2.19	12.4	18.0		33
Annual	42.8°	44.4°	31.79″	32.54″	68.5′	72.6′		51%

[a] Estimated.[2]
[b] Current normals.
[c] Includes melted snow.
[d] Percentage of total possible.
[e] Figures not available.

Late winter (February) is likely to display the greatest wintertime drop in temperature. That pattern of cooling is the same as that indicated by data from Boston, Massachusetts (table 24).

Backing up the implication of colder weather in the near future are the records of some bitter winter months in northern New England between 1880 and 1909. The coldest month ever at Hanover, New Hampshire, was January 1888, when the mercury averaged a mere 6.8 degrees. And the frigid winter (December, January, and February) of 1904-05 was the coldest on record (16.5 degrees) at Eastport, Maine. A winter such as that in coming decades could increase heating requirements by more than 15 percent over current normal demand. That is another way of saying

TABLE 24   Boston, Massachusetts, Monthly Averages

	Temperature		Precipitation [b]		Snow		Sunshine [c]	
	1880–1909 (1980–2009)?	1941–1970 [a]	1880–1909 (1980–2009)?	1941–1970 [a]	1880–1909 (1980–2009)?	1941–1970 [a]	1894–1903	1936–1976
Jan.	27.6°	29.2°	3.77′	3.69′	12.5′	12.6′	51%	54%
Feb.	28.1	30.4	3.47	3.54	12.3	12.5	57	56
Mar.	35.6	38.1	3.77	4.01	7.9	9.1	53	57
Apr.	46.2	48.6	3.14	3.49	1.8	0.7	53	57
May	57.0	58.6	3.48	3.47			57	58
June	66.3	68.0	2.76	3.19			60	63
July	71.5	73.3	3.06	2.74			60	66
Aug.	69.5	71.3	3.41	3.46			60	67
Sept.	63.5	64.5	3.38	3.16			62	64
Oct.	52.5	55.4	3.53	3.02			54	61
Nov.	42.3	45.2	3.28	4.51	2.1	0.9	45	51
Dec.	32.2	33.0	3.33	4.24	8.8	8.1	52	52
Annual	49.4°	51.3°	40.38′	42.52′	45.4′	43.9′	55%	60%

[a] Current normals.

[b] Includes melted snow.

[c] Percentage of total possible.

your fuel bills would be that much higher than average
— assuming the cost of fuel remains constant, which it will not, of
course.

Much of New England and upper New York State may be just a
bit drier in the near future, too. The Burlington numbers suggest a
drop in average annual precipitation of about 2½ percent. (The
Boston records imply a decrease of about 5 percent.) At Burlington
the biggest drying out is indicated for the month of April,
although the driest month ever was May. In May 1903 not enough
rain to measure fell. The driest year on record at Burlington was
1881, when just 20.99 inches of rain and melted snow fell. That
was almost 11 inches shy of normal.

Still, widespread drought should not be a problem. For instance, during the same year that Burlington was suffering from
lack of rain, Eastport, Maine, got drenched with 13.22 inches in
May, the wettest month in history there. Eastport's dampest year
ever came just a short while later. A total of 64.53 inches of
precipitation was measured in 1883.

Concomitant with the implied decrease in precipitation is a
suggested reduction in annual snowfall. Do not panic, skiers. The
magnitude of the reduction is not significant. As a matter of fact, a
closer examination of table 23 suggests the ski season might even
be lengthened a bit during the next few decades. With colder temperatures in the fall and spring, slightly greater November and
April snowfalls are likely. There are some records of the ski season
around Burlington getting off to a galloping start in November in
a number of years around the turn of the century. In November
1886, 30 inches of snow ushered in a season that saw a total of 142
inches come down. Next year (1887) saw November check in with
19 inches; November 1900 brought 2 feet, and November 1906 did
almost as well with 21.1 inches.

Even though seasonal snowfall totals might not average quite so
high in the immediate future as they did from 1941 through 1970,
that will not preclude some very snowy winters from occurring.

Eastport's snowiest season ever was 1906-07. A total fall of 187.5 inches had Down Easterners shoveling constantly that winter.

Still, there might be a few years that will turn out to be real bummers as far as natural snow is concerned. But not to worry. The implication of somewhat colder winter temperatures (especially during February) suggests that pretty good snow-making conditions will prevail most of the time. All in all, the picture seems a positive one for northern New England's ski industry.

Central New England

I have spent much of my professional life in southern and central New England and have grown quite fond of the area. But I have never been able to make a completely satisfactory adjustment to the long, cold winters, or to the high cost of living (especially around the Boston area). Unfortunately, for me anyhow, the weather records from 1880 through 1909 suggest that coming to terms with those two problems will not be made any easier in the next few decades.

The most significant implication of the turn-of-the-century records from Boston (table 24) is that winters in the immediate future are going to be longer and colder. Of the 10 coldest Boston winters since 1871-72, 5 occurred between 1880 and 1909. Overall, a drop in average annual temperature of around 2 degrees is suggested. Although that does not sound like much, it is significant on a yearly basis when averaged over 3 decades. What it means in terms of heating requirements is that they will increase by 10 to 15 percent a year. Or, putting it another way, we will probably see our fuel bills average that much higher in the near future. Remember, that outlook is based on an expected change in temperature alone. It does not take into account fuel cost increases that will undoubtedly continue to result from economic and political pressures. In the coldest winters, based on temperatures alone, fuel bills could jump 15 to 20 percent! The silver lining is that summers should be a bit cooler, too. That would mean less use of air conditioners.

There might not be quite so much sunshine in summers as there is now, but there is no hint that June, July, and August will be any wetter. In fact, the Boston records show that average yearly precipitation was about 2 inches less in the 1880–1909 period than it was during the 1941–1970 time frame.

Even though it might be slightly drier in coming decades, serious drought should not be a threat. The range of annual precipitation totals in the turn-of-the-century records was between 30.07 inches and 52.63 inches. That compares with a contemporary range of 23.71 inches to 62.32 inches. The implication is that rainfall and snowfall over the coming decades will show less annual variation.

No important change in seasonal snowfall averages is suggested, despite the fact that total precipitation may be somewhat reduced. There is a clue, however, that spring and autumn snowfalls will be more significant. For instance, in Boston, there is one case of 15 inches of snow falling in April (1887) and another of almost 18 inches in November (1898). At this point it is probably appropriate to note that both the latest and earliest freeze on record in Boston occurred in the early 1880s. Latest: May 3, 1882; earliest: October 5, 1881.

That the effective winter seasons will be longer in coming decades is also implied by the fact that Boston's October and November temperature averages of 1880–1909 show the biggest monthly departures from the 1941–1970 normals; March and April are runners-up. A look at the sunshine percentages in table 24 suggests that in addition to being chillier, October and November will also be cloudier. At last the good news: Little change in the amount of sunshine is implied for winter and spring.

Data from Blue Hill Observatory, located about 10 miles south of Boston and on top of a 640-foot hill, support the fact that New England winters were longer and colder around the turn of the century (i.e., the temperature trends evidenced in the Boston record have not been greatly influenced by urbanization). Winter season (December-January-February) temperatures averaged 25.7 degrees from 1880 through 1909. The contemporary normal is 27.5 degrees.

Another measure of the change in wintertime conditions is the average numbers of days between the freezing and thawing of nearby Houghton's Pond. The average was 112 days during 1887–1909, compared with 98 days for 1941–1970.

Since individual statistics such as temperature and snowfall do not always give a good feel for the severity of a winter, something called a *severity index* was developed for Blue Hill.[3] The severity index is defined simply as the total number of days per season in which the temperature fails to exceed 32 degrees plus the number of days per season with 6 or more inches of snow on the ground. For instance, in the winter of 1976-77 (one of the more severe winters in recent history in New England), the temperature failed to top freezing on 51 days, and there were 58 days that had 6 or more inches of snow on the ground. Thus, the severity index was 109.

Severity indices weren't tallied before 1895, but in the period 1895–1909 they averaged 70. The 1941–1970 average was 61. From 1963 through 1977 (to facilitate comparison with the initial 15-year period 1895-1909) the index averaged 65.

Grandpa was right, then. New England winters *were* tougher when he was a kid. Over the next few decades the rest of us may get a chance to find out what he was talking about.

New York City and Surrounding Areas, Including Southern New England

New York City and nearby areas will probably get to share in the tougher winters in the near future. The implications of the 1880–1909 figures from New York City (table 25) are that yearly average temperatures will drop about 2 degrees, average annual

TABLE 25   New York, New York, Monthly Averages

	Temperature		Precipitation [b]		Snow		Sunshine [c]	
	1880–1909 (1980–2009)?	1941–1970 [a]	1880–1909 (1980–2009)?	1941–1970 [a]	1880–1909 (1980–2009)?	1941–1970 [a]	1894–1903	1875–1976
Jan.	30.9°	32.2°	3.55″	2.71″	8.6′	6.8″	53%	50%
Feb.	30.6	33.4	3.75	2.92	9.1	7.5	58	55
Mar.	37.6	41.1	3.61	3.73	6.2	5.8	54	56
Apr.	49.1	52.1	3.15	3.30	0.7	0.7	57	59
May	61.0	62.3	3.39	3.47			56	61
June	70.5	71.6	3.23	2.96			60	64
July	75.3	76.6	4.37	3.68			57	65
Aug.	73.3	74.9	4.38	4.01			60	64
Sept.	67.3	68.4	3.80	3.27			60	63
Oct.	55.6	58.7	3.72	2.85			52	61
Nov.	44.7	47.4	3.07	3.76	1.8	0.4	51	52
Dec.	34.8	35.5	3.40	3.53	5.9	7.1	52	49
Annual	52.6°	54.5°	43.42″	40.19″	32.3′	28.3″	56%	59%

[a] Current normals.
[b] Includes melted snow.
[c] Percentage of total possible.

precipitation will increase about 8 percent, and mean seasonal snowfall will be greater by roughly 14 percent.

As with Boston, the implied cooler temperatures would suggest an average increase of 10 to 15 percent in heating fuel usage compared with the normal requirements between 1941 and 1970. In the coldest winters of the coming decades fuel bills could be 20 to 25 percent higher than those of a recent "average" winter. Again, this assumes no change in the cost of fuel.

Not only should winters be colder in coming decades, but most summers should not be quite so hot. In southern New England the summer of 1903 was the second coolest on record at New Haven, Connecticut—second only to the summer of 1816, "The Year

Without a Summer." Of course, slightly cooler summers would require less use of air conditioning. That in turn might mean a slight reduction in electric bills, helping to offset the higher cost of fuel in the winter.

Similar to the pattern of cooling suggested for northern and central New England, southern New England and regions around New York City should experience their biggest drop in average temperature in the early spring (March and April) and late autumn (October and November). February should show a good healthy dive in mean temperature, too: almost 3 degrees.

Colder falls and springs would suggest the risk of unusually early- and late-season frosts. That's exactly what happened at New Haven during the 1880s. The earliest freeze on official records came on September 30, 1888; the latest on May 30, 1884.

The turn-of-the-century records from New York City suggest not only lower temperatures in the near future but more snow, too: about a 4-inch increase in average annual snowfall over the next 20 to 30 years. If that does not sound like much, consider it in terms of one more 8-inch storm every other year. Maybe you should buy that snowblower after all.

No annual snowfall records were established in New York City during the 1880–1909 period. But an all-time monthly mark was set in March 1896, when a total of 30.5 inches smothered the city. There were some significant early-season snows, too. November of 1898 brought a total fall of 19 inches, and November 1882 checked in with 14 inches. Off the southern New England coast, Nantucket Island, south of Cape Cod, had a record snowy season in 1903-04 with a grand total of 82 inches.

As a snow event, the great "Blizzard of '88" remains unparalleled in modern history. Even the super snowstorm of February 1978, which shut down much of southeastern New England for almost a week, did not quite match the March 1888 blizzard. The 1888 storm lashed Connecticut, western Massachusetts, and southern parts of New Hampshire and Vermont with gale-force winds, near-zero cold, and immense snowfalls. Depths of 30 inches or more were common. Some totals reached as high as 50 inches! Middletown, Connecticut, in the Connecticut River valley, was buried under an unbelievable 50-inch fall. New Haven did almost as well with 46 inches.

We may not have to face another Blizzard of '88 in the near future, but there is a message implicit in the statistics we have looked at: New York City and surrounding areas eastward through southern New England are probably going to have to contend with more snow over the next several decades.

As a matter of fact, they are probably going to have to contend with more precipitation in general. The implication of the New York City data is that mean yearly precipitation will increase by about 8 percent, with the largest increases coming in February and October. (The Boston records also suggest a significant increase in October rainfall.) In contrast, November may show a noticeable decrease in average precipitation. (Again, the same implication comes from Boston's figures.)

There were indeed some damp periods between 1880 and 1909 in New York City and adjacent regions. The wettest month on record in the city was September 1882, when the Big Apple was washed with 16.85 inches of rain. That same month saw a state rainfall record established across the Hudson River in New Jersey. The skies over Paterson opened up with 25.98 inches. By the end of the year (1882) Paterson had also set a state mark for annual precipitation: 85.99 inches.

A couple of decades later a New York State record for yearly wetness was set. In 1903, 82.06 inches of water soaked Wappingers Falls (near Poughkeepsie) in southeastern New York State.

All in all, it seems as though good investments for people living in and around New York City and southern New England over the next 20 or 30 years would be heavier coats and sweaters, an extra snow shovel, and a couple of umbrellas (well, the wind invariably tears up one).

The Middle Atlantic States

The implication of the Washington, D.C., records (table 26) is that over the next few decades the Middle Atlantic states will show an even more extreme trend toward cooler, wetter, and snowier weather than will states to the north.

Average annual temperatures may decrease by over 2½ degrees, mean yearly precipitation may increase by 11 or 12 percent, and normal seasonal snowfall may be greater by a healthy 35 to 40 percent. Overall sunshine, according to the turn-of-the-century figures, probably will not change much.

The magnitude of the implied cooling at Washington may be somewhat overstated. This is because a great amount of ur-

TABLE 26   Washington, D.C., Monthly Averages

	Temperature		Precipitation b		Snow		Sunshine c	
	1880–1909 (1980–2009)?	1941–1970 a	1880–1909 (1980–2009)?	1941–1970 a	1888–1909 (1980–2009)?	1941–1970 a	1890–1903	1949–1976
Jan.	33.5°	35.6°	3.57″	2.62″	6.4″	4.9″	48%	48%
Feb.	34.5	37.3	3.81	2.45	8.0	4.7	50	52
Mar.	42.3	45.1	4.07	3.33	5.0	3.5	48	55
Apr.	53.0	56.4	3.18	2.86	0.4		53	57
May	64.1	66.2	4.10	3.68			55	58
June	72.2	74.6	4.11	3.48			63	64
July	76.3	78.7	4.57	4.12			64	62
Aug.	74.4	77.1	3.92	4.67			65	62
Sept.	68.5	70.6	3.35	3.08			68	62
Oct.	56.4	59.8	2.86	2.66			60	60
Nov.	45.6	48.0	2.36	2.90	1.2	0.7	51	53
Dec.	36.2	37.4	3.38	3.04	3.4	4.0	54	47
Annual	54.7°	57.3°	43.28″	38.89″	24.4″	17.8″	57%	57%

a Current normals.

b Includes melted snow.

c Percentage of total possible.

banization has taken place around the observation site. Urban growth leads to a "heat island" effect. The result around Washington may have been to force average temperatures in the 1941–1970 era to somewhat higher levels than would have been the case through strictly natural influences. This would produce a greater difference between the means of 1941–1970 and 1880–1909 than would otherwise be the case. Still, the natural difference is probably significant.

For instance, a look back at the 1880–1909 decades shows us that Washington's coldest winter on record came in 1904-05. The mean temperature for December, January, and February was just 29 degrees. And the coldest temperature ever observed in the capital area came during the great February 1899 cold wave: 15 below.

Other states in the Mid-Atlantic region also recorded some significant cold events around the turn of the century. In North Carolina both Raleigh and Wilmington set all-time temperature minima during the February 1899 Arctic blast. Raleigh registered 2 below and Wilmington 5 degrees. Another significant cold snap in the Tar Heel State came in December 1880. Record lows were chalked up at Cape Hatteras (8 degrees) and Charlotte (5 below). Norfolk, Virginia, tallied a record low of 2 degrees in February 1895. In Raleigh the coldest month ever experienced was January 1893 with a mean of 30.8 degrees.

Over the next few decades inhabitants of Washington, like the residents of Boston and New York City, will probably see their average heating bills hike 10 to 15 percent over what they've been in recent years. If a winter as frigid as 1904-05 were to be repeated, bills for December, January, and February would soar by over 25 percent. Again the caveat: These figures are based on potential temperatures and have nothing to do with changes in fuel costs. Any rise in the price of heating fuels would have to be added to the figures I have mentioned.

But the news is not all bad. Summers should not average quite so hot, and that would mean less work for air conditioners. For instance, in most summers in the near future, air conditioners might consume about 20 percent less electricity on the average than they have recently. That might help reduce utility bills, or at least soften the impact of any further increases in them.

Along with colder winters should come more snow, too. Annual totals around Washington may increase by 6 inches or more. Certainly in the 1880–1909 period there was a plethora of record snowfalls. Washington's snowiest winter ever was 1898-99, when 54.4 inches fell. A couple of other eastern cities set records that same winter. Trenton, New Jersey, measured 61 inches, and Philadelphia, 55.4 inches.

Raleigh, North Carolina, established a seasonal snowfall mark in the winter of 1892-93 with 31.6 inches. Lynchburg, Virginia, followed suit in 1895-96 with a total fall of 46.7 inches. And Scranton, Pennsylvania, recorded its snowiest winter in 1904-05: 88.6 inches.

Maximum depth records also piled up during 1880–1909. Washington's deepest snow (34.2 inches) was measured in February 1899, the same month the temperature plunged to 15 below. Numerous other cities also tallied all-time deep snows that month. Baltimore measured 30 inches on the ground; Atlantic City, New Jersey, 27.9 inches; Trenton, 30 inches; and Philadelphia, 26 inches.

Wilmington, Delaware, measured a record 22 inches on the ground in January 1909. Norfolk, Virginia's deepest snow was in December 1892: 18.6 inches. And Charlotte, North Carolina, had 16 inches on the ground in February 1902 to establish an all-time mark.

Not only more snow, but more rain, should plague the Mid-Atlantic states over the next 20 to 30 years. The records from Washington suggest that August and November could average a bit drier in the near future, but that the remainder of the months should all be damper. Reviewing the extremes from 1880–1909 reveals that Washington's rainiest year was 1889, when a grand total of 61.33 inches of precipitation soaked the area. Trenton also established a record for wetness that same year (1889) with 67.23 inches of rain and melted snow. In fact, 1889 was a record-breaking wet year all the way south through Virginia. Baltimore reported its soggiest year ever with 62.35 inches. And in Virginia Lynchburg (60.58 inches), Norfolk (70.72 inches), and Richmond (72.02 inches) all joined the parade. Charlotte, North Carolina, had jumped on the bandwagon a little earlier with 68.44 inches of precipitation in 1884.

Again, the implications for the Mid-Atlantic states are pretty strong: cooler, wetter, and snowier. Who knows, maybe politicians will become believers in the 180-year cycle, as well as in the 2-, 4-, and 6-year cycles.

The Southern and Central Appalachians

The 1880–1909 records from Knoxville, Tennessee (table 27), suggest that the southern Appalachians will share in the trend toward cooler average temperatures over the next few decades. However, the suggested cooling isn't nearly so great as that implied for other areas of the East. The central portions of the Appalachians may experience a greater decrease in temperature, reflective of the trends indicated by the figures from Detroit (table 16) and Washington, D.C. (table 26).

A slight increase in average annual precipitation in coming years is implied by Knoxville's records. But average yearly snowfall may diminish a bit. That is a suggested trend that could be questioned, however, in light of the possibility that winters will probably be slightly colder with greater average precipitation (table 27). Certainly, interpolation of the implications from Detroit's and Washington's data would indicate a significant increase in snowfall across the central Appalachians in the near future.

TABLE 27 Knoxville, Tennessee, Monthly Averages

	Temperature		Precipitation b		Snow		Sunshine e	
	1880–1909 (1980–2009)?	1941–1970 a	1880–1909 (1980–2009)?	1941–1970 c	1883–1909 (1980–2009)?	1941–1970 c	1898–1903	1943–1976
Jan.	38.7°	39.3°	4.66″	4.67″	2.8″ d	3.9″	41%	38%
Feb.	41.2	41.7	4.90	4.71	3.2 d	3.8	45	45
Mar.	49.2	49.1	5.47	4.86	1.2 d	2.1	46	51
Apr.	58.0	59.7	4.09	3.61	0.3 d		54	60
May	66.9	68.0	3.89	3.28			64	61
June	74.1	75.2	4.17	3.63			62	61
July	76.6	78.0	4.24	4.70			66	60
Aug.	75.8	77.1	3.89	3.24			67	61
Sept.	70.7	71.4	2.87	2.78			66	59
Oct.	58.8	60.4	2.39	2.67			67	60
Nov.	48.0	48.2	3.29	3.56	0.3	1.0	50	48
Dec.	39.9	40.4	3.98	4.47	2.1	2.3	46	37
Annual	58.2°	59.0°	47.84″	46.18″	9.9″	13.1″	56%	56%

a Actual, unadjusted means.
b Includes melted snow.
c Current normals.
d 1884–1909.
e Percentage of total possible.

The Knoxville figures point toward most of the cooling there over the next 20 or 30 years showing up in the months from April through October. Winters should be colder, too, but by less than a degree.

Even though the wintertime cooling may be relatively minimal, some truly frigid weather may make occasional visits to the region. Knoxville's coldest reading ever came in January 1884: 16 below. Farther north, Parkersburg, West Virginia, and Pittsburgh, Pennsylvania, established low marks in the famous February 1899 cold wave. Parkersburg slid to 27 below and Pittsburgh hit 20 below. In the central Appalachians Monterey, Virginia (near the border of West Virginia), set a state minimum temperature record, 29 below, in that same great Arctic outbreak.

The most frigid temperature ever observed in Pennsylvania came in January 1904. Smethport, in the northwestern part of the state, tumbled to 42 below.

The chillier winters may be accompanied by some deep snows from time to time, too. Knoxville's snowiest month ever was in February 1895: 25.7 inches. The greatest depth ever measured there was 22.5 inches in December 1886. Chattanooga, Tennessee, experienced its greatest monthly snow, 15.8 inches, in January 1893. The greatest depth there, as in Knoxville, occurred in December 1886: an even foot.

To the north, Pittsburgh's maximum monthly snowfall came in December 1890, when just over 3 feet (36.3 inches) fell on the city. Pennsylvania state snowfall records were established that same winter (1890-91): at Blue Knob (near Altoona) 86 inches—a monthly mark—fell in December, with the total accumulation for the season amounting to a record-shattering 225 inches. I guess it is not surprising there is a Blue Knob Ski Area.

The Knoxville records suggest that more rain will fall on the region over the next few decades. The increase in average precipitation may be most noticeable from March through June. October through December may actually experience a slight decrease. Still, annual totals may average 3 or 4 percent higher than contemporary normals.

Going back to the turn-of-the-century records, mention of a number of very wet events can be found. Knoxville's wettest month on record was January 1882, when 16.98 inches of rain and melted snow fell. In West Virginia Elkins's wettest month (11.0 inches) was July 1907; that year (1907) went on to become the rainiest and snowiest in history there: 65.37 inches. Parkersburg's dampest year was 1890, when 62.67 inches of precipitation soaked the town. In June 1901 Princeton caught 16.3 inches of rain to establish a West Virginia State record for maximum monthly precipitation.

In Pittsburgh July 1887 was the rainiest month on record: 9.51 inches. And 1890, with 50.61 inches of precipitation, was the soggiest year ever.

The Knoxville information on sunshine suggests that most Marches and Aprils, as well as being wetter in the near future, will probably be cloudier, too. But more sunshine should prevail from July through October, concomitant with less average rainfall for that same period.

Again, the message for the southern and central Appalachian region is pretty much the same as for the rest of the East: cooler, wetter, and, at least in the north, snowier over the next few decades. The southern Appalachians may see average heating requirements elevate by only 5 percent or so; but in the coldest of winters demand could jump by almost 25 percent. The central Appalachian area may well experience greater wintertime cooling than will the southern region. And that could result in average heating requirements running closer to 10 percent greater than recent normals.

The eastern United States as a whole should average significantly cooler during the next 20 to 30 years. At least that is the implication from the 1880–1909 figures. Analysis of a number of records suggest the biggest drop in temperature may be centered near Washington, D.C.

Mean yearly precipitation should increase over all of the East, except for northern and central New England, where a slight decrease is possible. The largest increase in precipitation, similar to the greatest decrease in temperature, may take place around Washington.

Average annual snowfall may be greater over most of the East, too. Slight reductions may occur in northern New England and in the southern Appalachians. However, the implied reduction for the southern Appalachians, as mentioned earlier in this chapter, can be questioned. Again, the largest increase in average seasonal snowfall may be in the area in and around our nation's capital.

The bottom line on all of this is that it is going to cost us more money. More money to heat our homes and higher taxes to help finance the cost of snow removal. Of course, that is happening anyhow, thanks to inflation. The probable change in weather patterns will just exacerbate the problem.

Small wonder so many people are moving to Florida and Arizona.

Notes

1. Schneider, S. and Mesirow, L. *The Genesis Strategy* (New York: Plenum Publishing Corp., 1976), pp. 42–43.
2. A crude correction factor was applied to Burlington's temperature readings between 1884 and 1906. The readings were taken by a co-operative (non-Weather Bureau) observer and appeared to be anomalously high. When the twenty annual means recorded by the co-op observer were compared with the twenty succeeding Weather Bureau-observed annual means, the co-op readings came out an average of 3.2 degrees higher. Means (all from the Weather Bureau) for the same 20-year periods in Boston and in New York City were compared. The 20-year period corresponding to the Burlington co-op observations averaged 1.2 degrees *lower* in Boston and .5 degrees *lower* in New York.

   An assumption was made that Burlington was under the influence of the same large-scale weather patterns during the period in question as were New York and Boston...that is, that the trends in annual mean temperatures should have been about the same at all three locations. Thus, choosing the more conservative of the differences from New York and Boston (New York's .5 degrees), a correction factor of −3.7 degrees (.5 degrees plus 3.2 degrees) was applied to all 1884–1906 monthly means at Burlington. Therefore, the 1884–1909 means must be considered as estimated.

   The resultant pattern and magnitude of differences between the monthly means of 1884–1909 and 1941–1970 matches very closely those calculated for Boston and New York. Therefore, the estimated temperatures at Burlington can probably be considered reasonably reliable.
3. Conover, J. "Are New England Winters Getting Milder?" II, *Weatherwise* 20 (1967): 58.

CHAPTER 12

The Weather of the Early 1800s: Implications for the Coming Decades

The New Haven Record

ONE OF THE longest continuous temperature records in the United States comes from New Haven, Connecticut. In 1780 Rev. Ezra Stiles began temperature observations in New Haven, and continued them until his death in 1795. At the time he was president of Yale College; the observations were continued at Yale until 1866.[1]

Certainly the records are not so comprehensive as modern ones, since the observations were taken only once or twice a day, usually at random hours.[2] And the exposure and location of the observations were different from those for modern readings. But for the purposes of comparing monthly means over 30-year periods, I will ignore those problems and suggest that we settle on considering the temperature differences between then and now as *approximate*. The rain and snowfall records are probably subject to less question, so they can be compared a bit more directly to contemporary figures. Unfortunately the precipitation records are not as complete as the temperature observations, but they still give us enough data to calculate averages over 20 years or so.

Basically the message from the New Haven records of 180 years ago is the same as that from the New York City records (and in part, the Boston records) of 100 years ago: The next few decades in southern New England are going to be colder, snowier, and wetter (wetter at least along the south coast).

So far we have been looking at weather records from 1880 through 1909 and considering them as indicative of conditions to come over the next several decades. Ideally, of course, it would be nice to have some definitive records from 180 years ago to use as a point of comparison. Remember, one of the primary cycles I have been discussing is a 180-year cycle. But, outside of some tree ring data in the western United States, and a tabulation of hurricanes along the East Coast, there is not much in the way of quantitative information available from the beginning of the nineteenth century.

Happily, though, some residents of Connecticut were quite weather conscious then and preserved some numbers for us. The Lewis and Clark expedition also left us some information. It is not much but it is enough to tell us what one winter in North Dakota was like in the early 1800s. But let us start with New England.

The pattern of future cooling implied by the temperature differences in the New Haven record (table 28) is very similar to that indicated by the Boston record (table 24). Both the New Haven and Boston data suggest that October and November will feature the greatest amount of cooling. And both records indicate winters will be on the order of a degree and a half colder; but the New Haven record shows the biggest wintertime difference in the month of January, not February, as do the New York City (table 25) and Boston figures.

The 3 earliest decades of the 1800s weren't particularly noted for record-setting, extreme cold, but the persistence of cold is worthy of mention. For instance, over a 13-year period at New Haven, from 1810-11 through 1822-23, there was only 1 winter that averaged milder than normal (based on the 1800–1829 figures), and that was by a mere 0.3 degree. The implications for our heating bills over the next 20 or 30 years are not very encouraging.

The New York City weather observations from 1880 through 1909 suggest an increase of almost 15 percent in annual snowfall for the coming decades. The older New Haven measurements imply the same thing. Average annual snowfall between 1803 and 1820 was 43.6 inches compared with a contemporary average of 37.1 inches—an 18-percent difference. Other snowfall ob-

TABLE 28   New Haven, Connecticut, Monthly Averages

	Temperature		Precipitation [c]		Snow	
	1800–1829	1941–1970 [a]	1804–1829 [b]	1941–1970 [a]	1803–1820	1944–1968 [a]
Jan.	26.1°	28.9°	3.51″	3.21″	15.2″	9.3″
Feb.	28.8	30.2	3.81	3.09	11.8	10.3
Mar.	35.8	37.4	3.43	3.97	7.5	6.9
Apr.	46.7	47.8	3.34	3.72	0.6	0.7
May	57.1	57.2	4.07	3.67	0.4	
June	67.2	66.8	3.30	2.73		
July	71.4	72.3	4.23	3.13		
Aug.	70.0	70.9	3.96	3.82		
Sept.	63.3	64.5	4.15	3.10		
Oct.	51.8	54.7	3.64	3.05		0.1
Nov.	40.5	44.0	3.74	4.25	1.6	1.1
Dec.	32.0	32.2	3.38	4.07	6.5	8.7
Annual	49.2°	50.6°	44.56″	41.81″	43.6′	37.1″

[a] Current normals.

[b] Record incomplete; 22 years of data for most months.

[c] Includes melted snow.

servations from the early nineteenth century confirm that that period was exceptionally snowy. Hartford, Connecticut, measured seasonal totals of 90 inches in 1820-21 and 85 inches in 1819-20. Both of those tallies would break the modern record of just under 83 inches set in the winter of 1966-67.

The winter of 1804-05 brought massive snows to the New York City area and much of Connecticut.[3] Almost 77 inches fell on New Haven that winter; that would surpass the contemporary mark of 76 inches in 1915-16. In Hartford the snow piled up 3 to 4 feet deep that same winter; the official record depth is 32.8 inches (February 1948). New York City was bombed with 5 feet of snow in just 2 months. December and January brought seven distinct snowstorms and 60 inches of snow to the city. The modern snowfall record for an entire season in New York City is just over 63 inches (1947-48).

At New Haven 43 inches of snow fell in January 1805. And at Hamden, just inland from New Haven, 57 inches came down, more than would be expected in an entire winter at that location. I wonder how modern, transportation-oriented New York City and Connecticut would cope with snows of those magnitudes. (I suppose they would have to call in consultants from Buffalo.)

(After the winter of 1822-23 snowfalls were relatively light for a 6-year period. At Hartford, the average annual snowfall for that 6-year stretch dropped off to 31 inches, compared with an 1817–1841 average of 42 inches.)[4]

The New Haven record closely matches the New York City record as far as implying that the next few decades are going to be wetter in those areas. Overall, the New Haven record suggests about a 6½-percent increase in average annual precipitation. The New York City figures indicate about an 8-percent jump. Even the pattern of month-to-month differences in the two records is virtually the same. That is, in months where the New York City precipitation differences imply an increase in the immediate future, the New Haven figures show the same thing. The only divergence of the differences is in the month of May. The New York City numbers suggest a slight decrease in May's average rainfall, whereas the New Haven figures imply an increase. Both sets of observations indicate a significant decrease of precipitation should come in November.

Climatologist Helmut E. Landsberg of the University of Maryland reconstructed a precipitation record for Cambridge, Massachusetts, a city adjacent to Boston. The reconstruction goes back to 1751, and indicates that during the period between 1800 and 1830 the average annual precipitation at Cambridge was about 40 inches.[5] This is virtually the same as was the 1880–1909 average annual precipitation at Boston (table 24).

Thus, both sets of figures support the contention that the decades of the immediate future will be slightly less rainy around Boston and perhaps in much of northern New England. As you have seen, data from New York City and New Haven suggest that the south coast of New England will be somewhat wetter during the same time.

It is interesting to note that the records from New Haven (1800–1829), New York City (1880–1909), and Boston (1880–1909) all point toward a noticeable change in late autumn weather over the next 2 or 3 decades. Octobers should be colder and wetter, Novembers colder and drier (but with snows showing up more frequently). The change implied for October is unfortunate because it has always been my favorite month in New England. After a long, muggy summer, the bright, mild days, and crisp nights of October always gave my spirits a lift. Now instead of showcasing the kaleidoscope of autumn foliage and making us forget the cold grayness of the coming months, October may serve as just a dreary gateway for the descent into a long winter.

The Lewis and Clark Expedition

Another place where the winter is long is North Dakota.

Meriwether Lewis and Captain William Clark spent the winter of 1804-05 at Fort Mandan, Dakota, Louisiana Territory. Lewis and Clark were leading an expedition up the Missouri River on their way to the Pacific Northwest. They wintered over on the Missouri, at Fort Mandan, which is just a few miles northwest of the present site of Bismarck, North Dakota.

Lewis kept a diary of the weather that winter,[6] the same winter that brought the deep snows to New York City and southern Connecticut. The December 18, 1804, entry in Lewis's diary read, in part: "The thermometer at sunrise was 32 degrees below 0. The Indians had invited us yesterday to join their chase to-day, but the seven men we sent returned in consequence of the cold, which was so severe last night that we were obliged to have the sentinel relieved every half hour." The previous day had seen the mercury at 45 degrees below zero, which would equal the coldest on record at Bismarck since 1875.

The frigid Dakota weather continued into January. The January 10 entry in Lewis's diary:

The night had been excessively cold, and this morning at sunrise the mercury stood at 40 degrees below zero. . . . A young Indian, about thirteen years of age . . . came in. . . . His father who came last night to inquire after him very anxiously, had sent him in the afternoon to the fort: he was overtaken by the night and was obliged to sleep on the snow with no covering except a pair of antelope skin moccasins and leggings and a buffalo-robe: his feet being frozen we put them into cold water, and gave him every attention in our power. About the same time an Indian who had also been missing returned to the fort, and although his dress was very thin, and he had slept on the snow without a fire, he had not suffered the slightest inconvenience. We have indeed observed that these Indians support the rigors of the season in a way which we had hitherto thought impossible.

Lewis knew when spring had arrived. March 29, 1805: "A variety of insects make their appearance, as flies, bugs, &c."

Notes

1. Ludlum, D. *Early American Winters, 1604–1820*, (Boston: American Meteorological Society, 1966), p. 76.
2. Personal correspondence with David Ludlum, April 1977.
3. Ludlum, D., op. cit. pp. 170–171.
4. _____. *Early American Winters, II 1821–1870*, (Boston: American Meteorological Society, 1968), p. 10.
5. Landsberg, H. "Two Centuries of New England Climate," *Weatherwise* 20 (1967): 52.
6. Ludlum, D. *Early American Winters, 1604–1820*, p. 252 ff.

CHAPTER 13

Cooler, Wetter, Snowier

ALTHOUGH THE WEATHER RECORDS of 80 to 100 years ago suggest that cooler and wetter conditions will prevail over much of the nation in the near future, such trends should by no means be nationwide. Figure 16 is a schematic of the temperature trends implied for the 3 winter months (December, January, and February) over the next several decades. The trends, of course, are the differences between the 30-year means expected for 1980–2009 and contemporary means (those of 1941–1970).

Cooler

Specifically, figure 16 indicates that colder winters will probably plague most of the United States east of the Rockies. An exception may be the middle-Mississippi Valley and the Old South. The greatest cooling is suggested for the Great Plains and for the Eastern Seaboard from Virginia to Massachusetts. In those areas, wintertime temperatures may average as much as 2 degrees colder than current normals. Even the Deep South can expect to be occasionally swept by some memorable cold outbreaks, as noted in chapter 10.

In the West, winters in Arizona and New Mexico should average only slightly cooler; but California winters could be as much as 2 degrees chillier. However, most of the West will probably experience very little temperature change in the winter months. The Great Basin area may even see some moderation in dark season readings.

Patterns of implied temperature change for the summertime (June, July, and August) are presented in figure 17. In general, the patterns are the same as those indicated in figure 16 for the winter. Only for the summer, the areas of little or no average temperature change may shrink. In the southeastern United States the region of minimum change will most likely concentrate in areas near the Gulf of Mexico and along the Atlantic coast south of North Carolina. Over the Far West the area of little change may

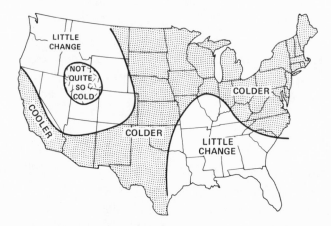

16. Wintertime (December, January, and February) average temperature conditions implied for 1980-2009, relative to those of 1941-1970 (current normals). "Little change" means averages are expected to change by only a half degree or less. The trend toward colder winters over the next several decades should be most noticeable over the Great Plains, and from Virginia northward into Massachusetts.

concentrate in the Pacific Northwest, northern California, and the Great Basin.

Maximum summertime cooling is suggested for southern California and the Southwest, and from the Great Lakes eastward into New England and the Middle Atlantic states. In those regions, summers could average as much as 2 to 3 degrees cooler (less hot) than contemporary ones.

On an annual basis, average temperature change patterns should be very similar to the wintertime patterns diagramed in figure 16. That is, yearly average temperatures will probably show little variation from contemporary normals in the Pacific Northwest and Great Basin areas, and in the middle-Mississippi Valley and Deep South. Elsewhere, noticeable cooling may prevail.

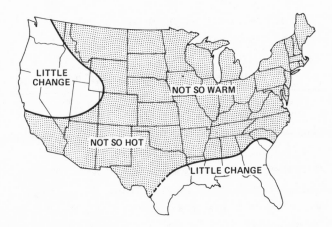

17. Summertime (June, July, and August) average temperature conditions implied for 1980-2009, relative to those of 1941-1970 (current normals). "Little change" means averages are expected to change by only a half degree or less. Maximum summertime cooling is suggested for southern California and the Southwest, and from the Great Lakes eastward into New England and the Middle Atlantic states.

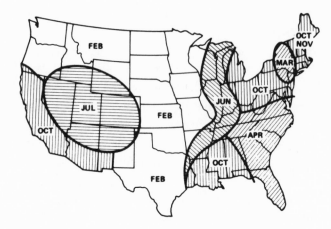

18. Months for which the average temperatures implied for 1980–2009 would show the greatest cooling relative to 1941–1970 means.

Figure 16 is interesting from another standpoint, too. Robert Kornasiewicz, a government climatologist, in the late 1960s investigated wintertime temperature patterns occurring over the United States in conjunction with certain weather types prevailing across Europe.[1] He identified the European weather types using a German catalog of *Grosswetterlagen* (large-scale weather patterns). One of the types defined in the catalog is a "westerly zonal flow, occurring farther south than usual": our old friend, the low-latitude westerlies. Kornasiewicz determined that in winters during which this type of flow prevails over Europe the temperature pattern in the United States is basically one of relative mildness in the West, and coldness in the East. The maximum positive temperature anomaly is centered in Utah, while the greatest cold anomaly—nearly twice the magnitude of the positive departure—extends from Minnesota into Illinois and then northeastward through Ohio.

The pattern indicated in figure 16, in a broad sense, is

remarkably similar to that. Thus, here is more evidence that the temperatures we have seen implied from the 1880–1909 normals have some real, physical relevance to low-latitude westerlies, the type of hemispheric upper-wind pattern Professor Willett hypothesizes will prevail for the next several decades.

In addition to considering seasonal temperature change patterns, it is interesting to contemplate the patterns produced by the months of potential maximum temperature change. Figure 18 is an example of what I am talking about. It is a map showing which month(s) will most likely display the greatest average cooling in various areas of the country. For instance, in the East, it is the spring and autumn months that will probably produce the largest drop in monthly mean temperatures over the next few decades. October and November may cool off the most from Ohio eastward and northeastward through New England. However, March will probably be the winner in eastern New York State.

In the Southeast, April may bring the largest negative departure from current normal temperatures. But then it is back to October for Louisiana, Mississippi, and Alabama.

West of the Mississippi River, winter and summer cooling will most likely dominate. A large portion of the United States encompassing the Pacific Northwest and the Great Plains will probably have to endure markedly colder Februarys in coming years. In the Interior West, July may bring the greatest cooling in average temperatures. The exception to the winter-summer pattern in the West could be California and southern Arizona, where October may offer the most significant drop in mean temperatures.

In between the overall eastern and western patterns, a strip of June maximum cooling will probably extend from the western Great Lakes southward toward the Gulf Coast.

Patterns produced by the months that may exhibit the least cooling (or greatest warming) can also be charted. Figure 19 indicates that the two most important months as far as minimum cooling (or most warming) goes will probably be September and March.

Over the next several decades, March will most likely feature a slight warming (relative to 1941–1970 normals) across much of the southern half of the United States. September may do the same thing across most of the northern part of the nation. In the intermountain West December could well offer the smallest degree of cooling; as a matter of fact, from Utah northward, December could average noticeably milder.

December will also probably produce the smallest drop in mean temperatures from the Ohio Valley eastward. No warming is likely, however.

In the coastal West various months from April through July should be the ones with minimum mean temperature change.

The coherent patterns displayed in figures 18 and 19 are something I did not expect to find when I began this project. I am

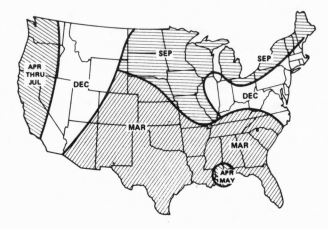

19. Months for which the average temperatures implied for 1980–2009 would show the least cooling (or in some cases, most warming) relative to 1941–1970 means.

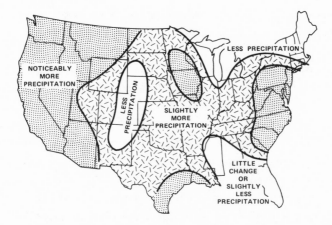

20. Average annual precipitation patterns implied for 1980–2009, relative to those of 1941–1970 (current normals). Except where noted, mean yearly precipitation over the United States should increase during the next several decades. The shaded areas represent implied increases in excess of 5 percent.

encouraged that they emerged, though. Their existence strongly suggests that the temperature changes we have discussed really are predominately the product of large-scale, atmospheric circulation changes, not the result of local influences. Local influences would be such things as observation site location changes, urbanization effects, and changes in thermometer exposures. The general pattern fostered by purely local determinants would probably be unorganized and random. For instance, the month of maximum cooling at Burlington might be May; at Boston, October; at New York, December; at Washington, August; at Detroit, March; and so on.

Wetter

Figure 20 displays the pattern of average annual precipitation changes, relative to contemporary normals, that may occur across the United States over the next 20 to 30 years. Most of the country should receive somewhat greater yearly precipitation, but a few areas may be a bit drier.

In particular, regions just to the east of the central Rocky Mountains may experience a noticeable drop in mean yearly precipitation. Slight decreases in average annual precipitation are likely from the Great Lakes eastward into northern and central New England. Small decreases may also occur in Mississippi, Alabama, and Florida, although that implication is open to question—see chapter 10.

However, major drought probably will not occur in any of those areas. The threat of severe drought should be highest in the southwestern United States, including the adjacent regions of southern California and western Texas; the best guess at timing would place the peak of the threat at around the year 2000. Chapter 3 presents a more detailed look at droughts.

Significant precipitation increases—greater than 5 percent on an annual average basis—will most likely be noted in the Far West, in Iowa and southern Minnesota, along the Texas and Louisiana Gulf Coasts, and from South Carolina northward through Pennsylvania and New Jersey to the New York City environs.

Snowier

The implications for changes in yearly average snowfall are a little unclear in some cases. But in general, more snow seems likely across the entire northern half of the United States, except for the western Great Lakes and northern New England. Both of those regions may experience a very slight decrease in mean annual snowfall.

The most significant trends toward less snow should come in the West. Areas in the lee of the central Rockies might have 10 to 20 percent less accumulation in most winters. And the lower elevations of the Great Basin region could see a decrease of similar magnitude. Higher elevations of the West will probably be blanketed by deeper snows in the majority of winters.

The regions likely to have to do battle with the largest increases in mean yearly snowfall over the next several decades are the Pacific Northwest, the eastern Great Lakes, and the Mid-Atlantic states.

When the areas of maximum cooling and maximum increase in snowfall are considered together, the one region they have in common is that centered around Washington, D.C. Thus, it is the Mid-Atlantic states that will perhaps have the most blatant trend toward harsher winters in the near future.

Some Reasons

Figure 21 is a schematic diagram of the type of upper-wind configuration that might prevail across the country for the next several decades. More precisely, it's a representation of what the

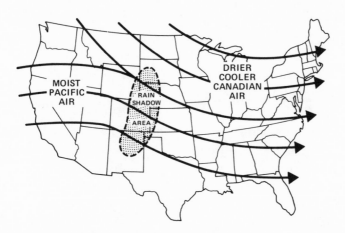

21. A schematic representation of the type of upper-wind pattern that might prevail, on the average, across the United States over the next 20 to 30 years.

pattern of the westerlies might look like if yearly patterns were averaged over the next 20 to 30 years. It would not be surprising if only a very few individual years exhibited a flow that looked anything like the average pattern. But, if all the annual patterns were lumped together, the resultant configuration would probably be very similar to that shown in figure 21.

The pattern of the westerlies indicated in figure 21 suggests a number of reasons for the trends of temperature and precipitation implied for 1980–2009. In the Far West the moist westerlies sweeping in from the Pacific Ocean would account for plenty of rainfall and little change in average annual temperatures. The small change in mean temperatures would probably result primarily from the failure of the westerlies to shift significantly farther south over the western United States.

Not only are the westerlies, on the average, unlikely to shift southward over the West, they may not move northward as often as they did during the past several decades. Meteorologists would say that the upper-wind pattern would have less *amplitude*. That means that a greater frequency of fronts and storms would charge directly inland, dumping more rain on the area, and producing moderate temperatures. In some seasons, such as the winter of 1977-78, which brought floods to Arizona and mudslides to southern California, the storm track could be unusually far south.

The fact that a greater number of fronts and storms, with more temperate Pacific air, may move into the West would account for the implied wintertime warming in the Great Basin area (figure 16). In the summer the same effect (more frequent invasions of temperate air) could be the reason that July may be the month showing the greatest amount of average cooling in the Interior West (figure 18).

During autumn the more frequent passages of fronts and storms through the intermountain West would prevent high-pressure cells from strengthening in that region. When high pressure does build up over the West in the autumn, the winds flowing off the high blow hot air into California and southern Arizona. In those areas, October is the month for which the greatest cooling is suggested in the near future (figure 18). Thus, the implication is that the westerlies will indeed prevent the development of strong high pressure over the Interior West; at least such development should occur much less often than in the recent past. That, for southern California, suggests fewer bouts with the dreaded hot, dry Santa Ana winds that whip brush fires through the region.

Farther east, as the westerlies cross the Rockies, the lee-side rain shadow effect (mentioned in chapter 6) should become apparent, especially in eastern Colorado. Annual precipitation totals could decrease by as much as 5 to 10 percent.

Over the eastern two thirds of the nation the westerlies (as diagramed in figure 21) should shift noticeably southward. Such a shift would permit the frequent intrusion of cooler, and in some cases drier, Canadian air into much of the region. The effects of the cooler air masses should be most obvious from the Great Plains eastward to New England and the Middle Atlantic states. Drier air from the interior of Canada should be dominant around the western Great Lakes and across much of New England, producing less average annual precipitation in those areas.

Immediately to the south of the Great Lakes and New England, more precipitation may fall in most years, especially along the Eastern Seaboard. This would be reflective of a southward shift in the primary storm path, following the trend of the westerlies to move south. It is this southward movement of the storm track that would likely be responsible for the significant increases in mean annual snowfall suggested for southern Michigan eastward through southern New England, and southward from there through the Mid-Atlantic states. The fact that many storms tend to reach their peak development as they move off the East Coast might account for the maximum increases in rainfall and snowfall implied for the Carolinas northward into southern New England.

As an example of a storm track that might be followed more

frequently in the near future, consider one which would run from Kentucky eastward to southern Virginia. To the south of the track increased shower and thunderstorm activity might occur as warm, subtropical air is drawn northward toward the centers of passing low-pressure areas. To the north of the path significant winter snowfalls might whiten the countryside as far north as southern Michigan and southern New England.

Of course, many different storm tracks will be followed over the years. The one mentioned here is given merely as an example of one that might occur more frequently in coming decades and that could account for the rainfall and snowfall increases implied by the turn-of-the-century data.

A southward shift in the main storm path would produce other effects besides changing precipitation patterns. For instance, such a shift would be consistent with the expected increase in the frequency of major tornado swarms in the Deep South in the near future. Tornadoes are most often spawned in squall lines generated by cold fronts, and cold fronts extend southward out of storm centers. Thus, the fact that the primary storm track might move farther south implies a concomitant shift of the severest twister activity.

Let us use the Kentucky-to-Virginia storm track as an example again. Cold fronts would trail southward from low-pressure centers racing eastward along that route. The fronts would sweep through Tennessee and the Gulf Coast states and then into Georgia and the Carolinas. Those are the states, then, that would experience the greatest threat of killer tornadoes.

The upper-wind pattern depicted in figure 21 gives an indication of why severe hurricanes might have a more difficult time reaching New England over the next several decades. Hurricanes approach the east coast of the United States from the tropical and subtropical Atlantic Ocean. That means they move from the south or southeast as they near our shores. The expected predominance in the near future of westerly winds at somewhat lower latitudes than during 1941–1970 would force many hurricanes eastward, away from the Atlantic coast. Only a very few of the more powerful tropical invaders might make it to American shores north of Cape Hatteras.

Our final item in regard to the westerlies is probably worthy of note before we move on. In the winter, a rather interesting pattern of temperature changes emerges when the means of 1880–1909 are matched against those of 1941–1970. The resultant implication for most of the country is that in coming winters December will show the least amount of average cooling (in some cases December may actually average milder), and February, the most.

Thus, the mean pattern of the westerlies will apparently shift only very slightly southward in December (where December averages milder, the westerlies might even edge a little northward), a little more so in January, and significantly more so in February. Particularly in the northeastern quadrant of the United States this may make February's cold indistinguishable from January's. At any rate, February, across much of the nation, will probably be the winter month displaying the most noticeable trend toward nippier temperatures.

What About All of the Recent Crazy Weather?

The winter of 1976-77 was the coldest in history in some parts of the United States. Unprecedented snowflakes fell as far south as Miami in January. In that same month, it was so warm in Alaska that bears, thinking it was spring, came out of hibernation. In the Far West an earth-cracking drought stalked California; and the Rockies and Cascades were virtually devoid of snow. In early May, a snowstorm worthy of mid-winter shocked residents of eastern Massachusetts. By July drought and record heat were making headlines in other parts of the nation.

At some locations temperatures soared to levels that had not been attained since the Dust Bowl days of the 1930s. Drought plagued farmers in the Southeast. But, in Johnstown, Pennsylvania, a deluge of rain triggered a flood that drowned at least seventy-six people. In August, an unusual visit by a tropical storm to southern California produced record rainfall. (It is perhaps a climatological commentary that the Los Angeles Dodgers, through 1977, had had only seven rain postponements since 1958. Six of those had come in 1976 and 1977.[2])

The wild weather continued into the winter of 1977-78. The drought in California turned into a nightmare of mud and floods as repeated heavy rains soaked the area. Meanwhile, awesome blizzards and roof-breaking snowfalls paralyzed various parts of the East at different times. A number of stations approached or surpassed maximum seasonal snowfall marks.

Many people are wondering, and rightly so, if this meteorological craziness is likely to persist. The answer is yes and no. (Well, what did you expect from a weatherman?) Yes, it could continue for another 5 or 10 years. No, it is not likely to linger through the next several decades.

The fact that such weather extremes may continue for another few years does not necessarily mean more record-shattering and snowy winters are in store for the East—although it could mean that. The winter weather blitz could shift into the West for a winter or two, then back to the East, and so on. Unfortunately, nobody can yet predict in which winters this might happen.

The recent extremes and variability of the weather are probably a function, at least in part, of a relatively large temperature gradient (contrast) that has been established between the mid-latitudes and tropical latitudes. The mid-latitudes have cooled off somewhat more than tropical and equatorial zones recently.[3] The resulting north-south temperature gradient has fostered some great atmospheric battles, with warm air punching northward and cold air southward, attempting to equalize the thermal imbalance. Such battles have been going on for eons, of course; they are part of the earth's "weather machine." But, because of the recent greater-than-normal temperature differences, these atmospheric wars have resulted in unusually harsh and greatly variable weather.

Professor Willett implies that this was to be expected, though. He points out that the influence of the long-term solar cycles in the tropics lags behind that in mid-latitudes by about 20 years.[4] Thus, the tropics probably did not reach their maximum warmth until the mid-1970s,* and probably will not reach their maximum coolness until early in the twenty-first century. In the meantime, the temperature gradient between the middle latitudes and tropics should begin to slacken. The result should be a trend toward less extreme and less variable weather. But that trend may take awhile to become established. Consequently for another 5 or 10 years weather events may still make frequent headlines.

Obviously, if you have read this book carefully, you will realize that there is one weather "extreme" that may persist well beyond 5 or 10 years. An increased frequency of notably harsh winters will probably linger for another 20 or 30 years.

Are We on the Threshold of an Ice Age?

The cooling of hemispheric temperatures in recent decades (see figure 9) has piqued interest in ice ages and the thought that we might rapidly be heading back toward such conditions. Since about 1965, however, the cooling has slowed considerably, and may even have stopped.[3, 6] I would expect any cessation of the cooling trend to be temporary, though; I would look for a resump-

---

*Willett feels that maximum hemispheric warmth wasn't attained until about 1953, well after the commonly accepted time of 1940 to 1945.[5]

tion of at least a gradual downward temperature trend that would persist for several decades. *

Some popular literature has cited the satellite detection of increased Northern Hemisphere snow and ice cover in 1971-72, 1972-73, and 1976-77 (a record cover occurred that winter[8]) as evidence of the "coming ice age." But the data are limited to about a decade. Consequently, there is no long-term base with which they can be compared. It is hard to make a plausible argument for an approaching ice age using only 10 years' worth of information.

The headline-making winter of 1976-77 is sometimes used as additional "evidence" that an ice age is just around the corner. Let me just note that the winter of 1976-77, on a *nationwide* scale, was only the fourth coldest in the past 100 years, not *the* coldest. The winters of 1898-99, 1909-10, and 1935-36 were all colder.[9]

But, we probably are, in fact, heading back toward an ice age! It will not be much of a problem for another 15,000 or 20,000 years, however, and this is discussed in the following chapter. It is runaway warming that may be much more of an immediate climatic threat, at least after the beginning of the next century; this, too, is discussed in the next chapter.

Books and articles on an imminent ice age make for exciting reading, but do not take them too seriously.

*In 1976 the main belt of Northern Hemisphere westerlies expanded to its largest size on record (such records were started in 1963).[7] That is another way of saying that, at least in some areas, the westerlies had shifted well south of normal. Perhaps this was a signal that the Northern Hemisphere cooling trend has already been reinitiated.

Notes

1. Wahl, E. and Lawson, T. "The Climate of the Midnineteenth Century United States Compared to the Current Normals," *Monthly Weather Review* 98 (1970): 259.
2. _____. "Weatherwatch," *Weatherwise* 30 (1977): 210.
3. Angell, J. and Korshover, J. "Estimate of the Global Change in Temperature, Surface to 100mb, Between 1958 and 1975," *Monthly Weather Review* 105 (1977): 375.

4. Willett, H. "Do Recent Climatic Fluctuations Portend an Imminent Ice Age?" *Geofisica Internacional* 14 (1974): 265.
5. Willett, H. and Prohaska, J. "Patterns, Possible Causes and Predictive Significance of Recent Climatic Trends of the Northern Hemisphere," The Solar Climatic Research Institute, Inc., 1977.
6. Kukla, G.; Angell, J.; Korshover, J.; Dronia, H.; Hoshiai, M.; Namias, J.; Rodewald, M.; Yamamoto, R.; and Iwashima, T. "New Data on Climatic Trends," *Nature* 270 (1977): 573.
7. Angell, J. and Korshover, J. "The Expanded North Circumpolar Vortex of 1976 and Winter of 1976-77, and Attendent Vortex Displacement," *Monthly Weather Review* 106 (1978): 137.
8. _____. "North American Record January Snow Cover," *Bulletin of the American Meteorological Society* 58 (1977): 439.
9. Quayle, R. and Diaz, H. "A Preliminary Note on the 1976-77 Winter Temperatures Over the Contiguous 48 States," National Climatic Center, 1977.

CHAPTER 14

Beyond the Year 2000

OVER THE PAST several million years the normal climatic condition of the earth has been that of an ice age, or what scientists call a *glacial* condition. Temperate, or *interglacial* periods, such as we currently live in, have existed only about 5 percent of the time.[1] Thus, we are living in a very unusual climatic period, at least compared with the past million years or so.

Cores of ocean floor sediment from the Pacific Ocean tell us that there have been seven major glacial eras over the past 700,000 years. These have alternated with warmer interglacials, which on the average have persisted between 8,000 and 12,000 years. The present interglacial has lasted about 10,800 years already.[2] This would seem to suggest that the long-term climatic trend should be back toward a glacial condition. Such a trend would be comparatively gradual, though,* and there would continue to be cycles superimposed upon cycles. That is, although a long-term

*The "snow blitz" theory, popularized on the Public Television presentation of "The Weather Machine" in 1975, holds that about 90,000 years ago glacial conditions developed in less than a century as heavy winter snows failed to melt during the summers. However, climatic conditions 90,000 years ago were such that the earth was already in a partial ice age.[3]

cooling trend might become established over the next several centuries, shorter fluctuations toward relatively warmer and colder temperatures would still be apparent.

Willett, for instance, forsees another climatic stress period occurring in the years 2110–2140. He feels it will be "substantially more severe than that of 1930–1960, probably as severe as that from 1370–1400."[4] During that period extreme cold and severe blizzards ravaged northern Europe in the winter, and blistering heat and drought plagued the region in the summer.[5]

After the climatic stress period, Willett feels a return to a "little Ice Age" is likely from about 2200 to 2550. This outlook is based on a 720-year cycle (four times the 180-year cycle). Though the validity of such outlooks can be argued endlessly, the validity of the forecast that we are going to be at least gradually heading back toward glacial times probably cannot be argued any longer.

The Milankovitch Theory

In the 1930s a Serbian geophysicist, Milutin Milankovitch, developed a theory that related global climate changes to the earth's orbital behavior. Basically, what Milankovitch did was postulate that the advance and retreat of ice ages were dependent upon variations in the seasonal and latitudinal distribution of solar radiation. He argued that these variations were caused by changes in the earth's orbital geometry.

Other researchers through the years either criticized or attempted to improve upon his work. But they, like Milankovitch, were hampered in their efforts by lack of an adequate global climate record covering a period long enough to prove or disprove their ideas. However, recent research by a team of American and British scientists has finally verified the Serbian's theory.

John D. Hays of Lamont-Doherty Geological Observatory, Palisades, New York, is head of the CLIMAP (Climate: Long-range Investigation, Mapping, and Prediction) project on which the team worked. Hays says, "We are certain now that changes in the earth's orbital geometry caused the ice ages. The evidence is so strong that other explanations must now be discarded or modified."[6]

By studying sediment cores from a relatively undisturbed section of the Indian Ocean floor, the CLIMAP project was able to uncover an unbroken geological record of climate covering more than 450,000 years. Thanks to modern dating techniques, this is three times longer than any previous chronology, at least one dated so accurately.[7] Close examination of the cores revealed significant climatic cycles of 23,000 years, 42,000 years, and 100,000 years—nearly the same as the overlapping cycles of variations in the precession and tilt of the earth's axis, and in its orbital eccentricity.[8]

The precession, or "wobble," of the earth's axis in its eliptical orbit around the sun means that over a period of centuries the earth is closer to the sun at different times of year. Currently the earth is closest to the sun in January. In about 10,000 years it will be closest in July. A complete precession has a cycle of about 21,000 years. Other things being equal, a greater summertime distance between the earth and sun should mean cooler temperatures and less melting of the polar ice caps (i.e., ultimate ice cap growth over a period of years).

The earth's orbit also tilts (in relation to the sun), and the amount of tilt varies over the centuries; thus the earth's axis is sometimes more perpendicular to the sun's radiation than it is at other times. The maximum deviation from perpendicular occurred around 9,000 years ago. Now the trend is toward a period of minimum tilt. A complete cycle of maximum to minimum and back to maximum tilt takes 41,000 years. Times of minimum tilt are conducive to ice ages. This is because at minimum tilt the earth's poles are more nearly equidistant from the sun all year, and there is a minimum of seasonal (i.e., winter to summer) effects. Much of the polar ice never melts and the ice caps slowly build. During periods of maximum tilt one pole, the one in the summer hemisphere, receives continuous solar radiation, and the ice caps tend to retreat.

The eccentricity of the earth's orbit also changes with time. That is, over a period of about 100,000 years it changes from nearly circular to more eliptical. The more eliptical path may act to accentuate the impact of the variations produced by the seasonal differences in distance between the earth and sun. For instance, when the orbit is more nearly circular, the precession effect (the 21,000-year cycle) has less influence because the seasonal differences in distance between the earth and sun are minimal. When the eccentricity is at a maximum, the seasonal differences produced by the precession are also at a peak. Thus, the eccentricity cycle may be the dominant of the three. The earth's orbit is now about halfway between the two extremes of the 100,000 cycle and is moving toward greater eccentricity.

Hays and his coworkers extrapolated the orbital cycles and put them into a climate model: "The results indicate that the long-term trend over the next 20,000 years is toward extensive Northern Hemisphere glaciation and cooler climate." They issue the caveat that the prediction ignores anthropogenic effects.[7] The problem is, we probably cannot ignore anthropogenic effects, those effects caused by man's pollution of the atmosphere.

Do We Live in a Greenhouse?

Carbon dioxide (CO_2) isn't normally considered an atmospheric pollutant. We cannot see CO_2, we cannot smell it, it does not damage our respiratory system, and it does not change the color of the sky.

Carbon dioxide is what we exhale when we breathe. Plants and trees like CO_2, of course, and use it in the photosynthesis process. That's the process that combines CO_2 and water in the presence of chlorophyll and sunlight to manufacture carbohydrates. All life depends upon photosynthesis either directly or indirectly. That is because all animals feed either on plants (carbohydrates) or on other animals that do. Obviously CO_2 is nice to have around. The trouble is, we are getting a lot more of it around now, and that might not be so nice.

Man's activities are injecting increasing amounts of CO_2 into the atmosphere. This comes about through the burning of fossil fuels—coal, oil, and natural gas. Fossil fuels are the product of the fossilized remains of plants and trees that lived millions of years ago. When we burn coal and oil, the CO_2 that was absorbed by plant life eons ago is released back into the atmosphere. But it is being released at a much faster rate than plant life can use it. The result is that CO_2 is building up in the atmosphere.

Scientists estimate that around 1860, when the industrial revolution began, atmospheric CO_2 had a concentration value of somewhere between 285 and 305 parts per million (or *ppm*—a chemical measure). Current estimates put the concentration at 330 ppm, an increase of between 8.2 and 13.8 percent since 1860, and of about 5 percent since 1958.[9]

Because the use of fossil fuels is continuing to increase, the amount of CO_2 in the atmosphere will also continue to increase. Researchers estimate that by the year 2000 the concentration value will be in the range of 380–400 ppm.[10] A recent study by the National Academy of Sciences (NAS) said that by the end of the twenty-second century, atmospheric CO_2 concentration might be four to eight times what it is now.[11]

"So what?" you might ask. "If CO_2 is not considered an atmospheric pollutant, why should we worry about an increasing amount of it?" Well, perhaps it should be classed as an atmospheric pollutant, because although it does not change the color of the sky, it may, by virtue of its increasing concentration, change the earth's climate. Carbon dioxide has the physical property of being relatively transparent to solar radiation (sunshine) but relatively opaque to long-wave radiation (heat). Or, saying it another way, CO_2 allows sunshine to heat the earth but then traps the heat in the earth's atmosphere, rather than permitting it to radiate back to space. This *greenhouse effect* warms the earth. And many scientists fear that the warming will accelerate as the demand for energy and thus the use of fossil fuels increases.

Current estimates suggest that CO_2-induced warming will account for about a 1.8°F rise in global temperature by the beginning of the twenty-first century.[10] Within a couple of hundred years, global warming could be on the order of 11°F with temperature increases in polar regions as much as three times greater.[11]

Dr. Wallace Broecker of the Lamont-Doherty Geological Observatory, one of the foremost authorities on the carbon dioxide issue, perhaps describes it better when he says the earth would be plunged into a "super-interglacial" with temperatures warmer than anything experienced in the last million years.[12]

Sea levels would rise and agricultural belts would be shifted by changing weather patterns. The NAS notes that "for some countries with marginal agriculture, the impact on food production could be severe."[11] If the warmth persisted over a number of centuries, the Greenland and Antarctic ice caps would melt, and oceans would rise, flooding the world's major coastal cities.[12] A recent popular novel treats the CO_2 problem in a fictionalized, much accelerated time frame, but the message is clear: Warmer is not necessarily nicer.[13]

To state the issue fairly, it must be pointed out that not all scientists agree with the carbon dioxide climate models. Willett, for instance, accurately points out that recent hemispheric cooling* has occurred during a period in which atmospheric CO_2 was in-

*See chapter 13 for further comments on recent hemispheric temperature trends.

creasing even faster than it was during the preceding hemispheric warming.[14]

By necessity, climate models are quite primative. We cannot stuff weather and climate into test tubes and perform laboratory experiments on them. The best that scientists can do is generate numerical mock-ups. The problem is that the mock-ups, or models, require various assumptions to be made. The assumptions may or may not be right. And the projected numbers that are plugged into the models may or may not be right.

For instance, in the case of the carbon dioxide problem, researchers can make a pretty good estimate of how much CO_2 is being released into the atmosphere every year. (They can do this by knowing how much fossil fuel is burned annually.) However, since 1958, only about half of the CO_2 known to have been produced by fossil fuel consumption in the same period has shown up in the atmosphere. Where has the rest of it gone? Scientists assumed for quite a while that it was being taken up by plants and trees, and by the oceans. But some biologists now maintain that none of the "missing" CO_2 is stored in plants.* And oceanographers say that though the oceans can absorb a substantial amount of CO_2, they cannot take up 50 percent of that produced by the burning of fossil fuels every year. Thus, things are going on that are not completely understood, and that makes modeling pretty difficult.

There are other difficulties too. For instance, there is the problem of "negative feedback." If increased CO_2 does, in fact, warm the atmosphere, evaporation may be increased and the result would be more cloudiness. More cloudiness would block more sunshine, and the warming effect of increased CO_2 could be

---

*Recent work indicates that through the destruction of the world's forests, land biota (plants and trees) may actually be acting as a net *source* of CO_2. The clearing of forests and the decay of humus may be adding as much CO_2 to the atmosphere annually as the burning of fossil fuels.[15]

negated.[16] Atmospheric cooling might even be triggered. There is "positive feedback" also. In that scenario CO_2-induced warming warms the oceans; because warmer oceans can retain less CO_2, they release previously absorbed CO_2 back into the atmosphere to further accelerate the warming.[17] Eventually, runaway heating of the earth results. Another possibility is that natural cooling cycles might override any CO_2 effects. Such natural cooling could produce cooler oceans, which, in turn, might be able to absorb even greater amounts of CO_2, thus reducing the whole problem to insignificance.[18]

Still, the consensus among most researchers in the field is that the carbon dioxide problem is real and presents a serious threat to the earth's climate. Although I feel that natural climatic cycles are still dominant, I also feel that this may not be the case much longer. By early next century the trek toward a super-interglacial may indeed be underway. The argument for diminished dependence on fossil fuels is obvious. But action must begin now. The transition to other energy sources will take decades to complete. If we wait until the CO_2 climate theory is verified (by observed global warming) or until improved models are developed, it will be too late. The CO_2-modeled warming is logarithmic (that is, more CO_2 leads to even faster warming), and once it is underway, there is probably no way to reverse it. We might successfully curtail it by switching, say, to solar energy, but much of the damage would have been done by then.

The models might be wrong, of course. But they could just as well be wrong in *under*predicting the amount of potential warming as in overpredicting it. Dr. Stephen H. Schneider, a climatologist at the National Center for Atmospheric Research, points out that "since the consequences of a climate change at the higher end of the current estimate could be both enormous and possibly irreversible, perhaps society would be best to err conservatively in planning future fossil fuel consumption patterns."[19]

J. Murray Mitchell of the U.S. Environmental Data Service

eloquently puts the problem into perspective: "Ours is the generation that may have to act, and act courageously, to phase out our accustomed reliance on fossil fuels before we have all the knowledge that we would like to have to feel such action is absolutely necessary. If we harbor any sense of responsibility toward preserving spaceship Earth, and toward the welfare of our progeny, we can scarcely afford to leave the carbon dioxide problem to the next generation."[12]

Six Hundred Million Tons of Dirt

Though most researchers feel that carbon dioxide offers the greatest threat to our climatic future, not all of them do. There are a few climatologists who think that anthropogenic particles (dust and smoke) have been and continue to be the primary culprits in changing our climate. Reid Bryson, director of the Institute of Environmental Studies, University of Wisconsin, is the leading proponent among those who adhere to that theory. Bryson and Gerald Dittberner of the University of Wisconsin developed a model that indicates that atmospheric dust is responsible for about 90 percent of the Northern Hemisphere's temperature variation this century.[20, 21]

Bryson argues that the 500 to 600 million metric tons of material injected into the atmosphere each year by man's activities act to reflect sunlight away from the earth. The average temperature north of the tropics is thus decreased, and the westerlies expand (shift to a lower latitude).[20] This, of course, is what Willett expects will happen over the next several decades, although he thinks the cause will be a change in solar radiation.

Bryson's assumption, or model, that anthropogenic dust in the atmosphere leads to cooling is not universally accepted. It is possible that under certain situations atmospheric particles may actually act to increase temperatures by absorbing long-wave (heat) radiation from the earth: the "greenhouse" effect again.[16]

This situation would be most likely to arise over land areas.[22] So, as with the CO_2-climate model, there are problems with the dust-climate model.

These problems are brought to the forefront when the pattern of recent hemispheric cooling is considered. Willett's work, for instance, indicates that land masses have been cooling off more rapidly than have the oceans, particularly the Pacific Ocean.[23] If particles do indeed lead to warming over land areas and are the primary cause of recent climate fluctuations, this pattern of cooling would not be expected.

Willett also points out that the most extreme cooling has been in the latitudes of 60°N to 80°N. That is north, he says, of where the most pronounced cooling induced by man-made dust would be expected.[23] (If anthropogenic dust were the main cause of the recent cooling trend, maximum effects should have shown up in the industrialized mid-latitudes, say between 35°N and 55°N.)

Willett goes on to state that the greatest cooling at high latitudes has occurred in the spring and autumn, as opposed to the summer and winter. That pattern, he argues, strongly implicates "solar corpuscular radiation" as a cause,[23] not dust, which should lead to a more seasonally uniform cooling. *

Thus, the major models dealing with man's influence on climate are fraught with problems. But this by no stretch of the imagination means they should be ignored. They are all we have to go on, and they are sounding a very serious warning: Our activities are going to change the earth's climate. Currently, the argument for CO_2-caused warming seems a bit stronger than other theories. But the ideas of Bryson and others should not be ignored.

Let me repeat one sentence: Our activities are going to change the earth's climate. That's particularly scary because scientists

---

*The discussion of the effects of solar corpuscular (charged particle) radiation (SCR) is beyond the scope of this book. I will do no more than mention briefly that SCR has been proposed as a mechanism for the control of atmospheric ozone, which in turn has a "greenhouse"-type effect on atmospheric temperature.

don't know for sure where that is going to lead after another 20 or 30 years.

I do not suppose we would be willing to exercise the political and social courage and make the economic sacrifices necessary to minimize the threat of man-made climate change. After all, mankind does not have a history of acting in advance of crises.

I fear, therefore, that my thought that we should greatly accelerate our effort to develop a safe, renewable, nonpolluting energy source—such as sunshine—is a little naive. Such a program would not only demonstrate foresightedness, but it would undoubtedly be costly and therefore unpopular. It would also deny us a lot of fun things to worry about. Such as running out of natural gas, oil embargoes, air pollution, outrageous utility bills, and, of course, warming (or cooling) our planet to the point of no return.

Notes

1. Libby, W. "Climatology Conference," *Science* 192 (1976).
2. Bryson, R. and Murray, T. *Climates of Hunger* (Madison: The University of Wisconsin Press, 1977), p. 131.
3. Schneider, S. and Mesirow, L. *The Genesis Strategy* (New York: Plenum Publishing Corp., 1976), p. 66.
4. Willett, H. "Do Recent Climatic Fluctuations Portend an Imminent Ice Age?" *Geofisica Internacional* 14 (1974): 265.
5. Ibid.; and Lamb, H. "On the Nature of Certain Climatic Epochs Which Differed From the Modern," *Proceedings of the WMO/UNESCO Rome (1961) Symposium on Climate Changes (Arid Zone XX)*, UNESCO (1963), 125.
6. _____. "Milankovitch Theory Verified," *Bulletin of the American Meteorological Society* 58 (1977): 184.
7. Hays, J.; Imbrie, J.; and Shackleton, N. "Variations in the Earth's Orbit: Pacemaker of the Ice Ages," *Science* 194 (1976): 1121.
8. Todd, E. Statement before the House Subcommittee on Science, Research, and Technology 8 February 1977, *Bulletin of the American Meteorological Society* 58 (1977): 521.
9. _____. "Rise in Global Concentrations of Carbon Dioxide," *Weatherwise* 30 (1977): 207.

10. These figures represent the general range of concentrations established by several studies during the 1970s; also see: Schneider, S. and Mesirow, L. *The Genesis Strategy* (New York: Plenum Publishing Corp., 1976), p. 179.

11. _____. "NAS Panel is Concerned over Atmospheric CO_2 Buildup," *Physics Today*, October 1977, p. 17.

12. Mitchell, J. "Carbon Dioxide and Future Climate," *EDS* (March 1977), p. 3.

13. Herzog, A. *Heat* (New York: Simon and Schuster, 1977).

14. Statement at American Meteorological Society meeting, Boston Chapter, October 1977.

15. Woodwell, G. "The Carbon Dioxide Question," *Scientific American* 238 (1978): 34.

16. Machata, L. "Mauna Loa and Global Trends in Air Quality," *Bulletin of the American Meteorological Society* 53 (1972): 402.

17. Flohn, H. "Climate and Energy: A Scenario to a 21st Century Problem," *Climatic Change* 1 (1977): 5.

18. Personal communication with H. P. Sleeper, Jr., Kentron International, January 1978.

19. Schneider, S. "On the Carbon Dioxide-Climate Confusion," *Journal of the Atmospheric Sciences* 32 (1975): 2060.

20. Bryson, R. and Murray, T. op cit. pp. 151–152.

21. _____. and Dittberner, G. "A Non-Equilibrium Model of Hemispheric Mean Surface Temperature," *Journal of the Atmospheric Sciences* 33 (1976): 2094.

22. Schneider, S. and Mesirow, L. op. cit. p. 181.

23. Willett, H. and Prohaska, J. "Patterns, Possible Causes and Predictive Significance of Recent Climatic Trends of the Northern Hemisphere," The Solar Climatic Research Institute, Inc., 1977.

CHAPTER 15

Personal Implications

WILL THE IMPENDING weather changes discussed in this book be serious enough for you to react to? For instance, should you consider moving from Massachusetts to Florida, or from Oregon to Arizona, or Minnesota to Texas?

Those are very personal and very important decisions. They are decisions whose resolutions are traditionally dominated by considerations other than those of climate change.

Ultimately, factors determining whether or not you change your place of residence are such things as job opportunities; availability of friends and relatives; and the general desirability of an area, based on an assessment of educational systems, medical facilities, crime rates, population densities, and shopping and recreational facilities.

The Appeal of the Sunbelt

Examination of weather factors will become increasingly important in the near future, however. Northern areas, relative to southern regions from southern California eastward through the Southwest and Texas into the Gulf Coast states, should become much more expensive places in which to live. Colder winters will create greater demand for heating fuel in the North, and that alone will raise the cost of living there. On top of that, increased snow removal problems and all of its associated costs will plague many northern cities.

Certainly among retirees, those factors themselves will continue to send multitudes streaming into the Sunbelt states, especially Arizona and Florida. Eventually, drought and water shortages may halt the immigration into Arizona. And, unless such projects as the Central Arizona Project, which will draw water from the Colorado River, can be completed, even residents already in place in Arizona could be faced with severe water rationing as a way of life.

Perhaps the southeastern states—the Carolinas, Georgia, Mississippi, Alabama, and Florida—will eventually become more attractive to retired people. Those states have greater rainfall than does the arid Southwest and, according to the U.S. Water Resources Council, may be among the few states that will have adequate water supplies by the year 2000.[1]

However, even in Florida water supplies could be threatened. Saltwater invasion of underground freshwater supplies may increase as more and more water is taken from subterranean reservoirs, allowing sea water to enter.[2] Thus, in Florida, as well as in Arizona, population growth may eventually have to be curtailed, or at least slowed.

But what about nonretirees? What if you are essentially locked into a particular northern location because of a job, or just simply because it is "home," what can you do? How should you react to the potential changes I have talked about?

Basically, I think it will boil down to consideration of a life-style change, not a geographical change. Frankly, many people would not consider a geographical change to a warmer climate even if they had the chance. In their own obscure, perverse way they like the changing seasons, the storms, the icy winds, and the snow. I know. I like them, too. (Each winter I like them a little less, though.)

Sweaters and Mukluks

But you are going to have to pay a price if you remain in the northern United States, particularly east of the Rockies, where significantly colder winters should prevail over the next several decades. Utility charges and heating fuel bills have already shot out of sight in many areas. I live in an all-electric condominium in northern Massachusetts. I was one of the few people in the development who were able to keep their utility bill for January 1978—which was only moderately cold—under two hundred dollars.

Many months in the near future will be even colder than January 1978. And in the meantime, fuel costs, and therefore electric costs, will continue to escalate. Within 10 years it may be a real challenge to keep my bills under three hundred dollars in winter months.

Millions of people around the country will face similar challenges. And, I suspect, they will be forced to change their way of living in order to combat rising heating costs, just as I have been forced to change mine.

Thermostats in most homes and buildings are kept at lower settings in the cold months now. That is a change that is here to stay, at least for a generation. Heavier shirts and sweaters will become more traditional indoor wear. A light long-sleeved sweater equals almost 2 degrees of added warmth. A heavy long-sleeved sweater is worth about 3.7 degrees. And two lightweight sweaters, because the air space between them acts as insulation, are equivalent to roughly 5 degrees of additional warmth.

Fifteen years ago in Point Barrow, Alaska, I had an Eskimo make a pair of mukluks for me. Mukluks are Eskimo "slippers" constructed from walrus hide and various animal furs. I never wore the mukluks much until the past couple of winters. They keep my feet warm in my cooler home better than anything I have. If you do not have a pair of mukluks, try wearing two pairs of socks.

Assuming that you have invested in adequate insulation, caulking, and storm windows, there are other adjustments you can make in your style of living to save money and at the same time make yourself more comfortable.

For instance, my wife and I live in a four-story, open-construction-type condominium. It is laid out vertically instead of horizontally. Depending on the season, we migrate upstairs or downstairs. In the coldest months of the year we essentially live on the two upper floors. We do this because—you should remember from your basic physics—heat rises. So we take advantage of that principle by adjusting our living to the most comfortable areas of our home. Instead of cranking up the heat to warm the lower levels, we surrender them to cooler temperatures. We vacate the living room in the winter, but there are only two of us. We do not really need a big living room.

In the summer, we shift our living to the lower floors. You may have noticed how cool your basement remains on the hottest days of the summer. As heat rises, cool air sinks. Luckily, we have a finished basement, so we decided to take advantage of the cool air down there during the hot months by buying a bed for the basement and sleeping there. That way we avoid running the central air conditioning full blower all night during heat waves just so we can sleep in our upstairs bedroom.

If you do not have the option of moving downstairs in the summer and up in the winter, there are other ways of minimizing heating and cooling costs. Basically, if you are willing and able, you will probably come out ahead by carrying on most of your activities in a few rooms instead of in an entire house. In the winter, you would be better off heating only a single room or two instead of a whole house, at least on the coldest days. If you do not have thermostats in each room, use space heaters to warm the area rather than keep a big central heating unit fired up.

By the same token, instead of running a central air-conditioning system during a hot spell, you could save electricity by operating window units in one or two rooms.

It is wonderfully relaxing and cozy to sit around a blazing fire in a fireplace, but it is a terribly inefficient method to heat a home or even a room. In a normal home fireplace, about 90 percent of the heat from the fire goes up the chimney. Not only that, but the fire sucks in air from the surrounding room(s). That air is eventually replaced by colder air seeping in from outside. Ultimately, you end up with a cold house—except maybe for the fireplace room—or the furnace kicking on to warm the remainder of the house while you enjoy your cozy fire.

I have got what is called a freestanding fireplace. It is made of metal and sits out away from the wall—thus, freestanding. The metal radiates heat very nicely and I have found it an effective way to heat my condo in the winter, providing it is not windy outside and the temperature is not below 25 degrees. There are not too many days in December, January, and February in northern Massachusetts that fit those criteria.

What I am saying is, if you have a fireplace, determine the limits within which it really is effective at warming a room or a living area. You may find you are paying an unnecessary penalty for being able to toast marshmallows. You might be better off with mukluks than a fireplace.

Using the Weather

You should pay attention to the outside elements, the weather, when you are trying to heat or cool your home. For instance, in the winter, by the simple expedient of keeping curtains closed on windows through which the sun is not shining you can prevent a good deal of heat loss. Keeping draperies drawn becomes especially important when a strong, cold wind is blowing directly on a window. And obviously, the heavier the curtains that you use, the more heat loss that can be prevented.

On the other hand, keeping curtains open when the winter sun is shining through a particular window can help warm the interior of your house. Providing a wind is not whipping directly against the window, this can provide effective additional warming even on a very cold day.

In certain sunny areas of the country, having more windows in a house might lower wintertime energy consumption. A University of New Mexico researcher, Wybe van der Meer, studied some randomly selected homes in Albuquerque. "In New Mexico," he says, "even single glazing on south, east and west sides of a house can be a net energy gainer, so long as you draw the drapes at night and open them during the day." He goes on to say that wall and roof colors are important. "In some cases, they may actually be working as low-grade solar collectors." Van der Meer points out that in Albuquerque if a light-colored wall with 2 inches of insulation were painted a dark color, the result would be "approximately equivalent to doubling the amount of insulating during the heating season."[3]

In the summer, when you are trying to keep your home cool, you want to keep sunshine out. That means closing the curtains when the sun is attempting to shine in. If there is a good breeze blowing outdoors, and you can cross-ventilate your house or apartment, open some windows to take advantage of the natural cooling effect of the wind. Of course, if it is oppressively hot outside, this might not be a good idea. But on most moderately hot days, it can be a lot more cost-effective than running an air conditioner.

It can be a bit of a pain in the neck to run around opening and closing curtains all day, but the result can be a more comfortable place in which to live, and lower utility costs. If you are away from home all day, you might want to carefully check the weather forecast before you leave. Depending upon the amount of sunshine and wind forecast and the predicted temperatures, you could choose which drapes to keep drawn and which ones to leave open. For example, consider a December day forecast to be cloudy and cold in the morning and then sunny with light winds in the afternoon. Before you leave home you might want to shut curtains on north- and east-facing windows but open ones on windows facing south and west in order to take advantage of the afternoon sun.

Perhaps someone will invent an automatically timed drape opener and closer. I think such a device would be a much better energy saver than a clock thermostat. My wife and I do not have any problem remembering to turn the thermostat down before we leave our home or go to bed. As a matter of fact, we constantly cross-check each other on that matter. For us, a clock, or time-of-day, thermostat would be a waste of money.

Of the energy you buy for your home, about 60 to 70 percent is used for heating and cooling, 20 to 30 percent is used to heat water, and 10 percent is used to run small appliances, lights, and stoves. So anything you can do to cut down on heating and cooling costs whittles away significant chunks of your utility bills.[4]

Take a Shower

Hot water is a necessity of American life. But many families probably have water that is too hot. Consider setting the thermostat on your water heater down to 120 to 130 degrees. Traditional settings are around 140 or 150 degrees. You probably do not need water that hot unless you have an automatic dishwasher. If that is the case, you can purchase an inexpensive inline heating unit for your dishwashing machine. By reducing the temperature of your water from 140 to 120 degrees, you can save over 18 percent of the energy you use at the higher temperature. You will not be taking cold showers at 120 degrees, because you never run the shower on full hot anyway. You always mix in some cold water.

By the way, taking a shower uses about half the water that a bath does. But if you have a wife who does not like to get her hair wet, you will probably never be able to enforce a complete, family life-style change on the matter of showers versus baths. One thing you might do is install a water-saving device on your shower. That would cut down on hot water usage, and water usage in general. It is possible that water-saving devices may eventually become mandatory in some areas, such as the southwestern United States, faced with severe water supply shortages.

Skis and Shovels

Besides wearing mukluks and sweaters around the house and taking showers instead of baths, I have made other concessions to cold weather—concessions you might consider. I decided instead of cursing New England winters from behind closed drapes, I would learn to coexist with them. Because I refuse to run in weather in which my breath can be seen, my jogging season in Massachusetts is limited to roughly April through October, though I can sometimes squeeze a November out of it. That used to mean that winters were virtually devoid of physical activity for me.

But I finally got tired of rotting away indoors and joined the snow lovers camp. A few years ago I purchased some cross-country ski equipment. I fell down a lot my first year on the trails, but now I actually look forward to snowstorms. With longer, snowier winters implied for much of the northland, I expect I will get plenty of use out of my skis over the coming years. The appeal of cross-country skiing should continue to increase: It is an inexpensive, easy-to-learn sport and a great way to beat the winter blahs.

Downhill skiing will not suffer in coming years, either. If anything, conditions should be even better in the near future than they were over the past few decades. Ski seasons should average longer in New England and snowier in the mountains of the West.

Colder winter temperatures in many areas may make outdoor ice skating more appealing. Snowmobiling, too, should become more popular. But I would really rather see more skiers than snowmobilers in the woods. Cross-country skiing is cheaper, healthier, and quieter.

While winter sports may boom over the next 20 to 30 years, summer sports certainly will not suffer. They should become more attractive, too, since cooler summers will likely prevail across much of the country and encourage all sorts of outdoor activities.

The projected changes in climate over the next few decades should be conducive to both winter and summer recreation, then. Use the changes to your advantage. Learn a winter sport, take up tennis or jogging in the summer: Work off the frustrations of higher heating costs, and get in shape for shoveling snow. Remember, in some regions there is going to be a lot more to shovel.

Perhaps if you live around the Washington, D.C., area, or in Portland, Oregon, you have been able to get away without a snow shovel or snow tires for many years now. But with large increases in average annual snowfall likely in those places in the near future, maybe it is time to think of buying a shovel, or finally breaking down and getting some snow tires.

Across most of the northern United States, where ice and snow will probably be much more assertive in coming decades, front-wheel-drive cars should become extremely popular. I have owned both front- and rear-wheel-drive autos, and on the snowy winter roads of New England, front-wheel-drive handling is far superior. I would think that a purchase of a front-wheel-drive car, if you live in a snowbelt area of the country, would be an excellent investment.

Maybe Doctor Bills Will Go Down

Finally, perhaps the most important question: How might the expected climate change affect your health in general? Assuming that you normally follow good health care practices, including doing such things as dressing properly for the weather, using a humidifier in your home in the winter—which is not only better for your health* but allows you to feel more comfortable with the thermostat on a lower setting—and pacing yourself when you shovel snow, cooler temperatures should make you healthier! A number of studies have determined that cooler, moderately variable weather is better for long-term health and more conducive to sharper mental performance.[5]

Ellsworth Huntington, of Yale University and author of *Mainsprings of Civilization*, found that the death rate of people under forty-five years of age in the cooler northern United States is generally half that of the same age group in the South. The lowest rates prevail from New England and New Jersey westward past the Great Lakes, then through the Great Plains from the Dakotas to Kansas. There is a break in the Rocky Mountains, but low rates are again dominant in the Pacific Northwest.

The North-South difference in death rates, some scientists think, is at least partially attributable to differences in metabolism. Metabolism refers to the chemical processes in living organisms that convert food and oxygen into tissue and energy. A by-product of metabolism is heat. For instance, human metabolism produces a body temperature of around 98.6 degrees.

When the outside air temperature rises to a point where losing body heat becomes a problem, your metabolic rate slows down to curtail heat production. If your metabolism becomes less active, so do other bodily functions, including your capacity to fight off infections.

Hotter weather may dull you intellectually, too. For instance, more people flunk civil service exams in July and August than at any other time of the year. The lowest marks in college tests are received in the summer.

Other studies have shown moderately variable weather to be more healthful and stimulating than other types of weather regimes. Uniform climates with little day-to-day variation—even

*If you wake up on winter mornings with a dry, scratchy throat, the humidity may be too low; using a humidifier would alleviate the problem.

long spells of bright, sunny conditions—seem to foster less endurance and less productivity in people. Extreme weather variations, including severe storms and rapid temperature changes, give you more bouts with such illnesses as colds and asthma.

All considered, the implied trend toward cooler, wetter weather across much of the country over the next several decades should have a positive effect on both your physical health and your mental alertness. In the southern United States, where cooler summers are suggested, the trend should be especially welcome. The general increase in precipitation likely across much of the country indicates that storms and fronts may appear at more frequent intervals. Thus, greater variability in the weather is probable. That, too, should keep you healthier and more active, other things being equal.

The weather conditions of the next 20 or 30 years may cost you more money. But if you learn to live with them, you may end up living longer because of them.

Cold Spring and Sun City

In the end, the personal implications of the climate change expected over the next 2 or 3 decades are perhaps more important than the meteorological changes themselves. Across most of the northern half of the country, what these personal implications boil down to is that you are going to have to learn to live with lower temperatures. You are going to have to learn to winterize your home and cars. You are going to have to learn to dress effectively for colder weather. You might even want to learn a winter sport. In short, you are going to have to come to terms with longer, colder, and snowier winters.

In other areas of the country the problems will be slightly different. Southeastern states bordering the Atlantic Ocean may be brushed by hurricanes more frequently. The Deep South should have warning and disaster-response systems in readiness for major tornado outbreaks. In the winter, the South should be prepared to contend with infrequent but significant snowstorms and cold waves.

In the West farmers on the High Plains adjacent to the central Rockies may have to deal with diminished precipitation. In the Southwest, severe drought may eventually halt immigration and economic growth. Water supply problems have to be solved now. In the Pacific Northwest water supply should be no problem, but because of wetter winters floods may be.

The most important impacts of the climate change I have discussed in this book will not be financial or health-related. The ultimate impacts will be related to the way you live.

If you can make a commitment to change your life-style slightly and to prepare for the changes I have talked about, then you'll be just as happy in Cold Spring, Minnesota, as in Sun City, Arizona.

Notes

1. "Is U.S. Running Out of Water?" *U.S. News & World Report*, 18 July 1977, p. 33.
2. Gustaitis, R. "Water," *Boston Sunday Globe*, 22 January 1978, p. A3.
3. _____. "Two New Reports on Insulation, Storm Windows Seemingly in Disagreement," *Solar Energy Intelligence Report* 3 (1977): 295.
4. Many of the ideas and facts presented in this chapter were obtained from pamphlets and publications furnished by the U.S. Department of Energy and Boston Edison, and through personal conversations with Eric Dickman, President, Energy Efficiency Corporation.
5. Everhart, M. "How Weather Affects Us or Looks Like We'd Best Be Cold!" *1978 Old Farmers Almanac*, 1978, p. 130; and Wolkomir, R. "Weather: The Powerful Influence It Has on All of Us," *Family Weekly*, 3 April 1977, p. 12.

Index

Elkins, West Virginia, 138
Elko, Nevada, 81
Embarras, Wisconsin, 96
energy
  need for non-polluting source, 177
  use in home, 185
Eureka, California, 59, 61

Fargo, North Dakota, 87
fireplace, inefficiency of, 182
Fishers Island, New York, 45
Fort Smith, Arkansas, 104
Fort Worth, Texas, 89–90
fossil fuels, need for diminished
  dependence on, 174
Fresno, California, 59
Fujita, T. Theodore, cyclical pattern
  of tornado occurrence, 49–50,
  108

Galveston, Texas, 109, 111
  as representative of climatic trend
    in deep South, 113–21
Galway, Joseph P., study of major
  tornado outbreaks, 51
Grand Island, Nebraska, 87
Grand Junction, Colorado, 80
Grand Rapids, Michigan, 96
Great Falls, Montana, 78
Great Salt Lake, 80
Green Bay, Wisconsin, 95, 97
greenhouse effect, 170–75, 175–76
Greenland, colonies on, as climatic
  record, 11
Greenville, South Carolina, 117
Grosswetterlagen, 152

Hamden, Connecticut, 144
Hanover, New Hampshire, 125
Hartford, Connecticut, 144
Havre, Montana, as representative
  of climatic trend in eastern
  Washington, eastern Oregon,
  Idaho, Montana, and
  northwestern Wyoming, 77–9
Hays, John D.
  head of CLIMAP, 168–70
  trend toward hemispheric
    glaciation, 170
health, effect of climatic trend on,
  187–89
heat island effect
  definition, 29
  in Los Angeles, California, 63

in Washington, D. C., 133–34
heating, cost of affected by
  climatic trend
  in Bismarck, North Dakota, 88
  in central New England, 127
  around Great Lakes, 99
  in Minneapolis-St. Paul, 88
  in New York City, 130
  in northern New England, 125–26
  in southern and central
    Appalachians, 139
  on southern plains, 90
  in Washington, D. C., 134
Helena, Montana, 78
H̄ouston, Texas, 111, 121
Huntington, Ellsworth, *Mainsprings
  of Civilization*, 188
hurricanes, 41–8
  Agnes, 47–8
  Camille, 47, 110
  in deep South, 109–10
  1893 storm in Louisiana, 109
  1893 storm in South Carolina, 110
  frequency in 1930s, 20
  Great September Gale, 45
  precautions for, 47
  recurvature path, 41

Ice Age
  normal climatic condition of earth,
    167
  threat in the future, 163–64
ice core records, 7, 9
Indianapolis, Indiana, 95–6, 97
interglacial periods, 167
  possibility of super-interglacial, 172
International Falls, Minnesota, 86

Jackson, Mississippi, 112
Jacksonville, Florida, 111
  as representative of climatic trend
    in deep South, 113–21
jet stream
  definition, 17
  supplier of energy to tornadoes, 51
Johnstown, Pennsylvania, 1977 flood,
  162

Kalispell, Montana, 78
Kansas City, Missouri, 90
Key West, Florida, 119
Knoxville, Tennessee, as
  representative of climatic trend
    in southern and central